Fundamentals and Applications of Bionic Drilling Fluids

Fundamentals and Applications of Bionic Drilling Fluids

GUANCHENG JIANG
State Key Laboratory of Petroleum Resources and Prospecting, China University of Petroleum, Beijing, China
MOE Key Laboratory of Petroleum Engineering, China University of Petroleum, Beijing, China

石油工业出版社有限公司
PETROLEUM INDUSTRY PRESS

ELSEVIER

Gulf Professional Publishing
An imprint of Elsevier

Gulf Professional Publishing is an imprint of Elsevier
50 Hampshire Street, 5th Floor, Cambridge, MA 02139, United States
The Boulevard, Langford Lane, Kidlington, Oxford, OX5 1GB, United Kingdom

British Library Cataloguing-in-Publication Data
A catalogue record for this book is available from the British Library.

Library of Congress Cataloging-in-Publication Data
A catalog record for this book is available from the Library of Congress.

ISBN: 978-0-323-90293-9

For Information on all Gulf Professional Publishing publications visit our website at https://www.elsevier.com/books-and-journals

Publisher: Glyn Jones
Editorial Project Manager: Naomi Robertson
Production Project Manager: Kamesh Ramajogi
Cover Designer: Christian J. Bilbow

Typeset by MPS Limited, Chennai, India

Contents

About the author

Guancheng Jiang, born in February 1966, earned a bachelor's degree in engineering from Southwest Petroleum University in 1987, a master's degree in engineering with a major of applied chemistry from China University of Petroleum in 1993, and a PhD degree in science with a major of marine chemistry from Ocean University of China in 2005. He is now a professor at China University of Petroleum (Beijing).

Professor Jiang has long been engaged in scientific research and teaching in the field of petroleum engineering. His main research interests include drilling and completion fluids, petrophysics, oil and gas field chemistry, and protection of oil and gas reservoirs. Professor Jiang's significant achievements in these fields represent the highest levels of scientific and technological research in China. His achievements in the field of polymer drilling fluids and protection of oil and gas reservoirs are among the most important around the world.

Professor Jiang won the National Science Fund for Distinguished Young Scholars of China, which is the highest honorary title for researchers engaged in basic theoretical research in China. The winner is commonly recognized as the candidate of the Chinese academicians with the highest academic status. He also won the Project of China's Thousands of Talents and China's Young and Middle-Aged Experts with Outstanding Contributions and was elected as an academician of the Russian Academy of Natural Sciences and the Russian Academy of Engineering in 2014.

Professor Jiang is the chief expert of the National High-Tech R&D Program of China (863 Program), the leader of the major special scientific and technological project in the oil and gas field during China's 13th Five-Year Plan (the only project leader in Chinese universities in the field of oil and gas), the deputy director of the Standing Committee of the Youth Work Committee of the Chinese Petroleum Society, and an editorial board member and chief reviewer of more than 10 scientific journals in the world.

Professor Jiang won the State Technological Invention Award once and the State Scientific and Technological Progress Award twice (awarded by the People's Republic of China for global researchers who have made outstanding contributions). He has received more than 30 provincial and ministerial awards for his scientific and technological achievements.

He has published 322 scientific research papers in the field of petroleum engineering in Chinese and English, including 104 SCI papers and 145 EI papers. He holds ownership of 70 Chinese invention patents, 6 Chinese computer software copyrights, 27 US invention patents, and 3 European invention patents. He has published seven academic monographs in China. In 2018 he was sponsored by Elsevier to publish the academic monograph *Gas Wettability of Reservoir Rock Surfaces with Porous Media* (ISBN: 978-0-12-815150-1), which has become the teaching material for senior undergraduates and graduate students at Chinese petroleum institutes.

Professor Jiang presides over 10 major scientific and technological projects in China each year, with annual research funding of more than 15 million CNY. Apart from this, he has supervised nearly 100 masters and doctors who, after graduation, have been working in various important scientific research positions in China's petroleum industry and have made significant contributions to the development of China's petroleum industry.

CHAPTER 1

New technology of borehole-enhanced bionic drilling fluids

The problem of borehole instability is a worldwide engineering and technical problem that is commonly encountered in oil and gas drilling and has not been completely resolved. According to the destruction strength equation of shale established by Eric van Oort, only by combining the three methods of improving the bond strength of shale, inhibiting clay hydration, and plugging the micro-nano pores of shale can borehole stability be maintained to the greatest extent. Since traditional drilling fluid additives cannot improve the bond strength of shale, Professor Guancheng Jiang of China University of Petroleum (Beijing) led his team to introduce bionics to the field of drilling fluids for the first time. By imitating the chemical structure of the marine mussel protein with superstrong underwater adhesion properties, a bionic wellbore strengthener that can effectively improve the bonding strength of shale has been developed, along with a bionic shale inhibitor that can inhibit the hydration dispersion of shale, especially shale surface hydration. By imitating the structure of bacteria, a bionic micro-nano plugging agent has been developed.

On this basis, with a bionic wellbore strengthener, a bionic inhibitor, and a bionic plugging agent as the core, in combination with other drilling fluid additives, a new theory, new methods, and new technology of wellbore-enhanced bionic water base drilling fluids featuring the synergistic effect of solidification, inhibition, plugging have been developed. As a result, the three major technical problems that were not solved by the previous anticollapse water base drilling fluid technology can be solved by the new technology. First, the combination of internal and external factors improves the bonding strength and cohesion between rock particles. Second, the pressure transmission in micro-nano pores is prevented. Third, the occurrence of osmotic hydration and surface hydration is inhibited. The indoor performance evaluation shows that the developed technology not only can inhibit the hydration dispersion of shale to a great extent, but also effectively strengthens the rock of the borehole, greatly improving the borehole stability. The field tests conducted at the

Fundamentals and Applications of Bionic Drilling Fluids
DOI: https://doi.org/10.1016/B978-0-323-90293-9.00001-0

three horizontal wells in Su 53 block of Sulige gas field in China, as well as North China Oilfield, Xinjiang Oilfield, and Sichuan Oilfield, show that this technology not only can effectively solve the problem of borehole instability in large sections of shale formations, but also significantly speeds up the overall development of unconventional oil and gas formations. Clearly, this technology has become a mature new drilling fluid technology.

1.1 Analysis of the characteristics of the rocks cemented by the secretions of marine mussels and bacteria that plug pores

1.1.1 Characteristics of the rocks cemented by the secretions of marine mussels and their bionic applications in drilling fluids

Mussels are a type of crustacean commonly found in coastal and offshore areas, especially in cold waters. They can form strong filaments of mussel adhesion protein secreted by their foot silk glands. They are often attached to the coastal reefs or the bottom of a ship and can remain firmly attached to the ship despite the scouring action of rough waves. Moreover, they can attach themselves very strongly to surfaces of metal, glass, polymer, minerals, and even Teflon (Fig. 1.1) [1]. In addition, the mussel adhesion protein has good biocompatibility and degradability and is a kind of bioadhesive with great advantages and potential.

Research has shown that the high-strength, high-toughness, and water-resistance properties of marine mussel adhesion protein as well as their extremely strong adhesion to substrates are mainly related to their

Figure 1.1 Mussels adhered on the surface of Teflon.

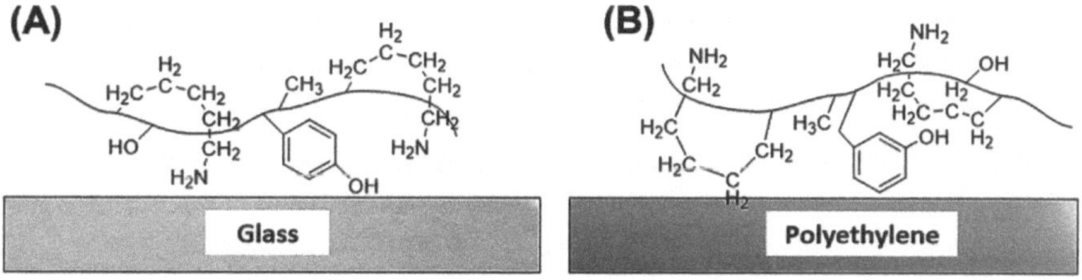

Figure 1.2 Schematic diagram of the adsorption behavior of mussel adhesion protein on (A) high-surface-energy substrate (glass) and (B) low-surface-energy substrate (high-density polyethylene).

special molecular structure, that is, dopa-mediated interchain cross-linking, and the interaction with substrate. With high affinity and chemical versatility, dopa can realize the adsorption and adhesion of different materials including organic matters and metals. When the mussel adhesion protein interacts with a high-surface-energy substrate (such as glass), amino groups and hydrocarbon groups in the protein molecule will react with the substrate to achieve infiltration (see Fig. 1.2A). For a low-surface-energy substrate (such as polyethylene plastic) the methyl groups in protein molecules and other groups will react with the substrate to achieve affinity (Fig. 1.2B). In this process, the phenol group of tyrosine residues shows a special affinity, and its affinity diversity can be adjusted by conformational changes [2]. When it interacts with a high-surface-energy substrate, a conformation in which the phenol hydrocarbon group is close to the substrate and the benzene ring is distant from the substrate is adopted. When it reacts with a low-surface-energy substrate, a conformation in which the benzene ring is close to the substrate and the phenol hydrocarbon group is distant from the substrate is adopted. The phenol group of tyrosine can be converted into the catechol group of dopa under the conditions of enzyme catalysis. Mussel adhesion proteins can achieve affinity diversity through catechol conformational changes, thereby enhancing the adsorption and adhesion of mussel adhesion proteins to different substrates [3,4].

The remarkable adhesion properties of mussel adhesion proteins make it suitable for unlimited potential uses in the fields of biomedicine, defense, and marine engineering. For many years, mussel adhesion materials have been prepared by traditional protein extraction methods and recombinant gene cloning methods, but these methods all have the disadvantages of excessive cost, and mass production and promotion of the materials are difficult. Therefore researchers began to shift their attention to low-cost ordinary synthetic polymers. In recent years, researchers have used DOPA/dopamine and mussel core adhesion functional elements $-COOH$ and $-NH_2$ as reactive functional groups. Through esterification or acylation reaction methods, they grafted them to synthetic high molecular chain with reactive functional groups or grafted them to the polymerizable monomer and then polymerize. This method obtains a polymer containing dopa or dopamine (DA), which replicates the effect of natural mussel adhesion protein. Research in this field can be divided into two categories: the functionalization of polymers and the polymerization of functional compounds [5–7]. China University of Petroleum (Beijing) developed a bionic wellbore strengthener that has been used in drilling fluids for the first time. It has a synthetic polymer bionic effect, and the bionic inhibitor is a direct modification product of mussel adhesion protein.

1.1.2 Characteristics of bacteria that plug pore throats and their bionic applications in drilling fluids

The structure of bacteria has a certain effect on their survival, pathogenicity, and immunity. The structure of bacteria can be roughly divided into surface structure, including the cell wall, cell membrane, and capsule; internal structure, including the cytoplasm, ribosome, nucleoplasm, plasmid, and spore; and external accessories, including the flagella and fimbriae. It is customary to categorize a structure that is indispensable for the survival of a bacterium or a structure that a general bacterium possesses as a basic structure, while a unique structure formed by certain bacteria under certain conditions is categorized as a special structure [8–10].

Because bacteria have the above-mentioned special structure, can be connected by peptide bridges or peptide chains, and can be cross-linked into a mechanically strong network structure, they have a resident and plugging effect at the pore throat or pore canal. This characteristic of bacteria has great reference value for the plugging technology of drilling fluids. The structural bionic method can be used to synthesize a bionic micro-nano plugging agent with a bacterial structure and special characteristics to prevent pressure transmission and stabilize the borehole.

1.2 Development and evaluation of a bionic shale inhibitor

The most common shale inhibitors used on site are KCl, quaternary ammonium salt-based small-molecule inhibitors, and amino polymer–based polymer inhibitors. Despite the successful application of these traditional inhibitors, they still have many defects. For example, KCl will cause serious damage to the environment and ecosystems, quaternary ammonium salt-based small-molecule inhibitors have only a little impact on the environment but still have a certain negative impact, the inhibition performance of amino polymers depends heavily on the pH value of drilling fluid, and the inhibition effect is weakened at high temperature.

In response to these problems, the bionic shale inhibitor of marine mussels, that is, DA, was synthesized, and the shale inhibition performance of DA was systematically evaluated through experiments involving linear expansion, hot rolling recovery, and bentonite interlayer distance. In addition, the mechanism of the action of DA was analyzed by means of infrared spectroscopy, X-ray photoelectron spectroscopy (XPS), zeta potential measurement, and so on, and the temperature resistance of DA and its influence on the rheological properties of drilling fluid were investigated. Finally, since DA can polymerize on the surface of shale pores to form a polydopamine (pDA) membrane, the shale-plugging effect of the pDA membrane was evaluated.

1.2.1 Preparation and characterization of bionic shale inhibitor DA

1.2.1.1 Preparation of bionic shale inhibitor DA

The synthesis of bionic shale inhibitor DA generally involves four steps: condensation, reduction, demethylation, and salt formation. The mechanism is shown in Fig. 1.3. First, vanillin is used as a raw material in a condensation reaction with nitromethane to form 4-hydroxy-3-methoxy-β-nitroethylbenzene (nitro compound for short). The nitro compound is reduced with zinc and hydrochloric acid as the reducing agent to obtain 4-hydroxy-3-methoxyphenethylamine salt (reduced product for short). The reduced product is hydrolyzed to remove the methyl group by hydrobromic acid and then converted to hydrochloride to obtain the final product of DA.

1.2.1.2 Characterization of DA

1.2.1.2.1 Infrared spectroscopy analysis

The infrared spectrum of DA is shown in Fig. 1.4. The stretching vibration peaks at 3348, 3230, and 3070 cm^{-1} correspond to N–H, O–H, and aromatic ring C–H, respectively. The peaks at 1616, 1600, and 1585 cm^{-1} are the asymmetric stretching vibration peaks of aromatic ring

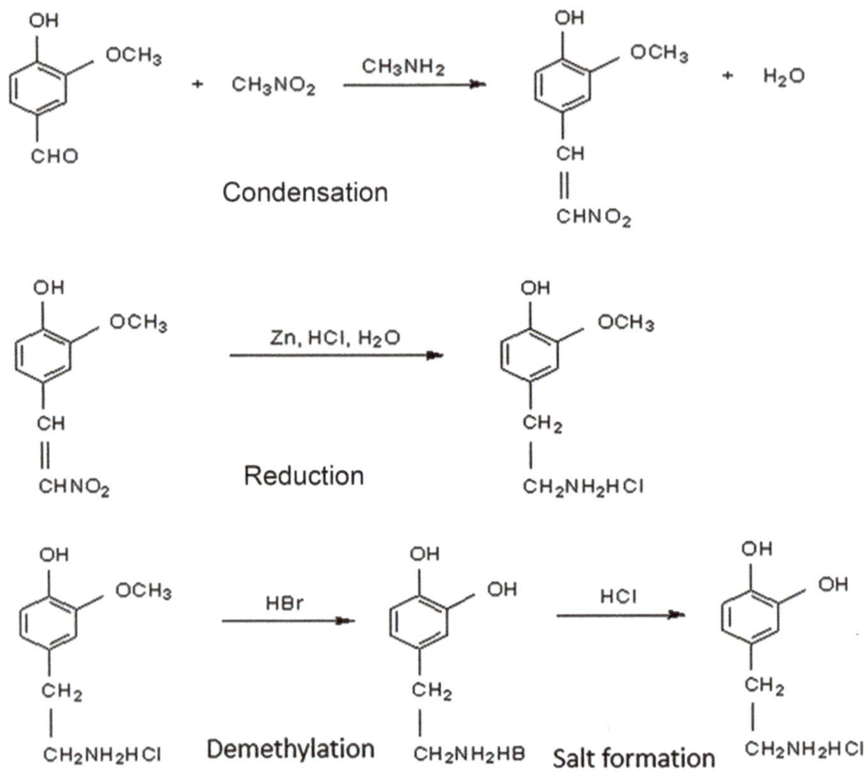

Figure 1.3 Synthesis mechanism diagram of DA.

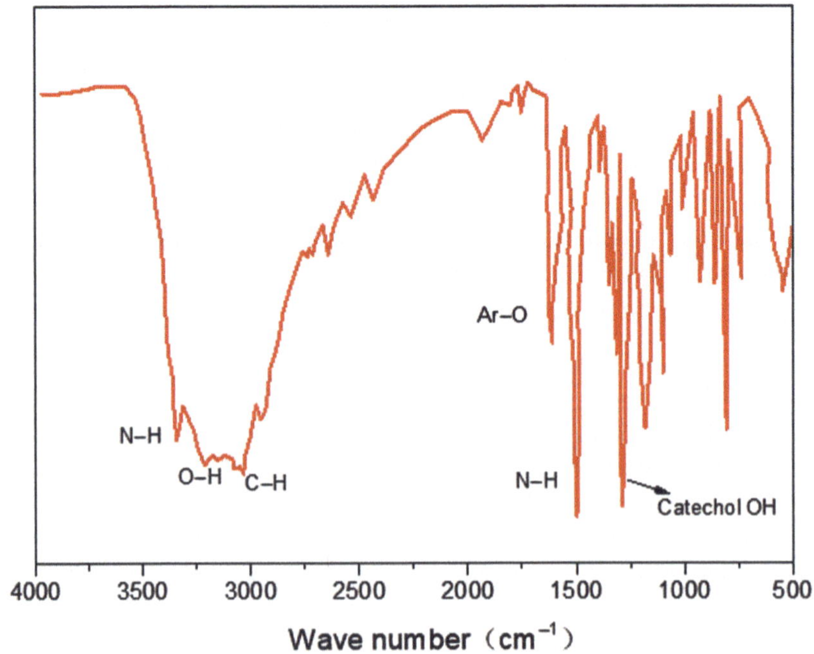

Figure 1.4 Infrared spectrum of DA.

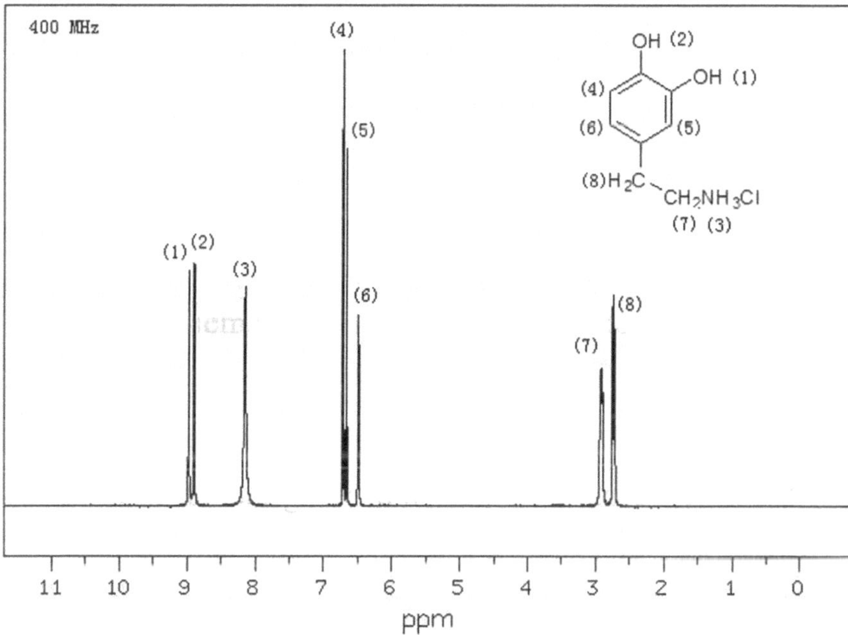

Figure 1.5 H-NMR spectrum of DA.

C—O. The peak at 1519 cm^{-1} is the bending vibration peak of N—H. The peaks at 1248 and 1281 cm^{-1} are the bending vibration peaks of catechol O—H. These results prove that DA has been successfully synthesized.

1.2.1.2.2 Nuclear magnetic resonance (NMR) analysis

The hydrogen nuclear magnetic resonance (H-NMR) spectrum of DA is shown in Fig. 1.5. The proportion of hydrogen in the spectrum is consistent with the standard product of DA, indicating that the product synthesized in the laboratory is of high purity.

1.2.2 Performance evaluation of DA

The inhibition performance of DA was systematically evaluated through the experiments of linear expansion, hot rolling recovery, and clay mud making. The bentonite used in the experiment conforms to the Chinese standard GB/T 20973-2007; the cuttings used in the linear expansion experiment are the 12,850-ft. shale in the San Francisco area, and the cuttings used in the rolling recovery experiment are the 15,330-ft. shale in the Villeta area, Colombia (similarly hereinafter).

1.2.2.1 Linear expansion experiment

Fig. 1.6 is the linear expansion curve of bentonite in different concentrations of DA solution (pH is adjusted to 8.5 with 0.1 mol/L NaOH). It can be seen from Fig. 1.6A that compared with pure water, the expansion amount of bentonite immersed in the DA solution is at a low level at any time period, indicating that DA has shale inhibition. As the DA concentration increases, the linear expansion of bentonite decreases significantly. When the concentration exceeds 3%, the expansion curve has leveled out within the experimental time. This shows that when the concentration reaches more than 3%, DA can be saturated and adsorbed on the surface of bentonite. Fig. 1.6B shows that within the initial 2 h, the expansion rate of bentonite in the 5% DA solution is actually greater than that in the 1.5% and 3% DA solutions. The cause of this abnormal phenomenon will be discussed in the section on inhibition mechanism analysis of DA.

Furthermore, through the linear expansion experiment of shale cuttings in 1% concentration of different inhibitor solutions, a comparison was conducted on the performance of current common inhibitors, including KCl, small cations (NW-1), and polyamines (Ultrahib of MI company), as shown in Fig. 1.7. Except for pure water, the time inflection point of the expansion curve of bentonite in each inhibitor solution is similar, indicating that the concentration of each inhibitor solution is much higher than the saturated adsorption concentration on the surface of bentonite. In this case, the expansion rate of bentonite in the DA solution

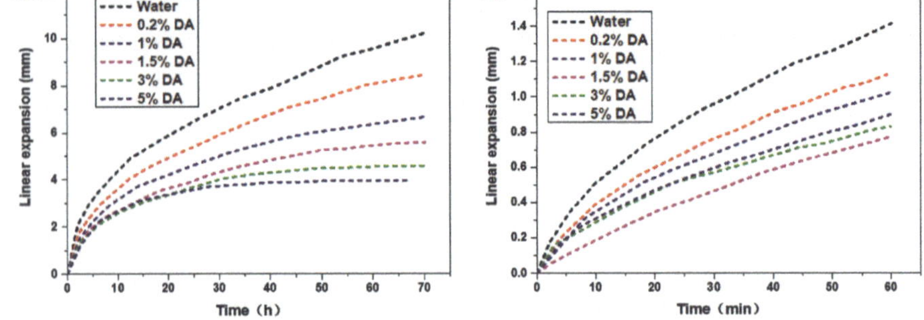

Figure 1.6 Linear expansion curves of bentonite in pure water and DA solutions at different concentrations after (A) 70 h and (B) 1 h.

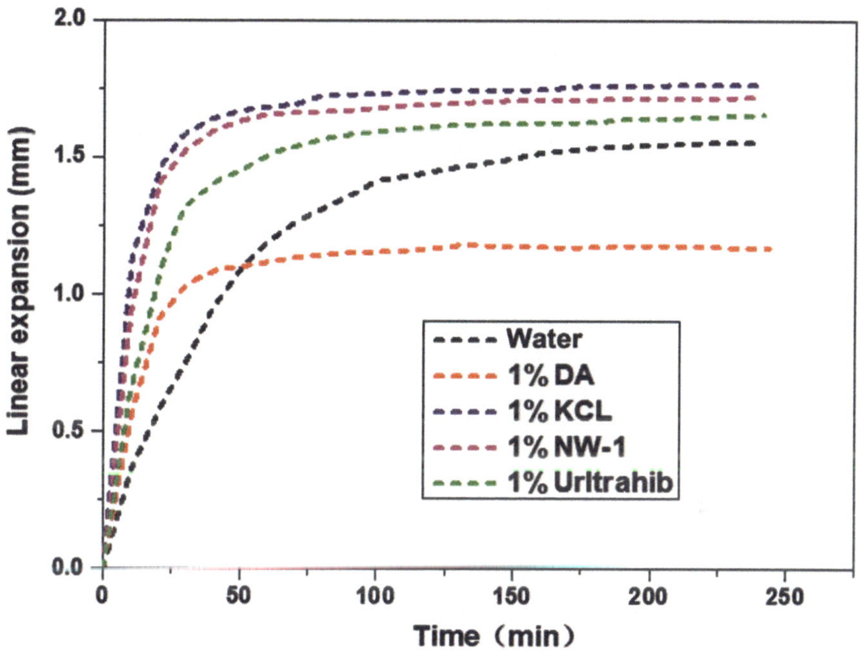

Figure 1.7 Linear expansion curves of shale cuttings in pure water and different inhibitor solutions with 1% concentration.

is the lowest in any period of time, indicating that the inhibitory effect of the DA is stronger than that of the others.

1.2.2.2 Hot rolling recovery experiment

The results of the hot rolling recovery experiment of shale in pure water and different inhibitor solutions are shown in Table 1.1. After hot rolling of shale cuttings in pure water at 120°C for 16 h, the recovery is only 43.05%. By contrast, the recovery of cuttings in 1% DA solution reaches 84.04%, which is significantly higher than that of other inhibitor solutions of the same concentration and is consistent with the results of the linear expansion experiment of shale, further proving that the DA has excellent shale inhibition performance [11,12].

1.2.2.3 Clay mud making experiment

The bentonite was added to the shale inhibitor solution and hot rolled at 150°C for 16 h. By measuring the change in the yield point (YP) of the inhibitor/bentonite dispersion system with the amount of bentonite added, the inhibition performance of the inhibitor on clay mud making

Table 1.1 Hot rolling recovery of shale cuttings in pure water and different inhibitor solutions with 1% concentration.

Types of inhibitors	Rolling recovery (%)
Pure water	43.05
0.2% DA solution	73.58
1% DA solution	84.04
1.5% DA solution	87.26
1% NW-1 solution	49.08
1% KCl solution	54.04
1% Ultrahib solution	77.62

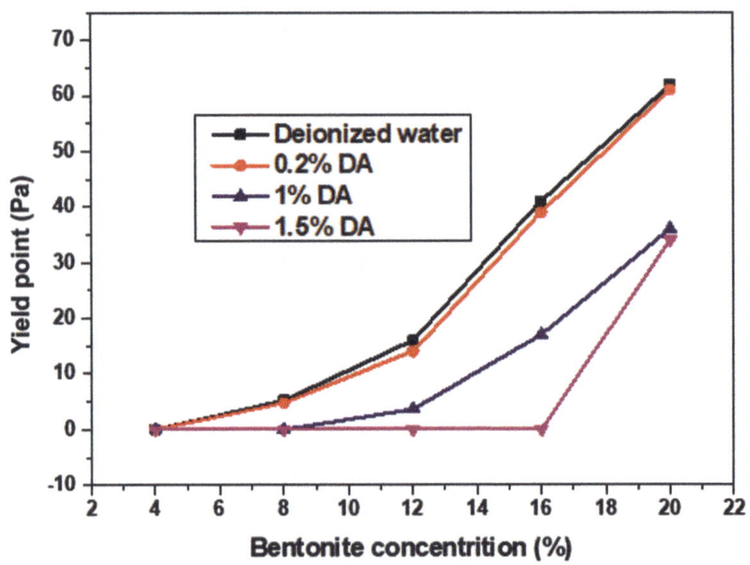

Figure 1.8 Change in YP of DA/bentonite dispersion with amount of bentonite added.

can be effectively evaluated. The YP of the bentonite dispersion liquid reflects the strength of the three-dimensional (3D) network structure formed by the bentonite particles during laminar flow. The lower the YP of the dispersion liquid, the weaker the network structure, and the stronger the inhibitory effect of the inhibitor.

Fig. 1.8 shows the change of YP of DA/bentonite dispersion system (pH ≈ 8.5) with the amount of bentonite. As the concentration of bentonite increases, the YP gap between samples with different DA

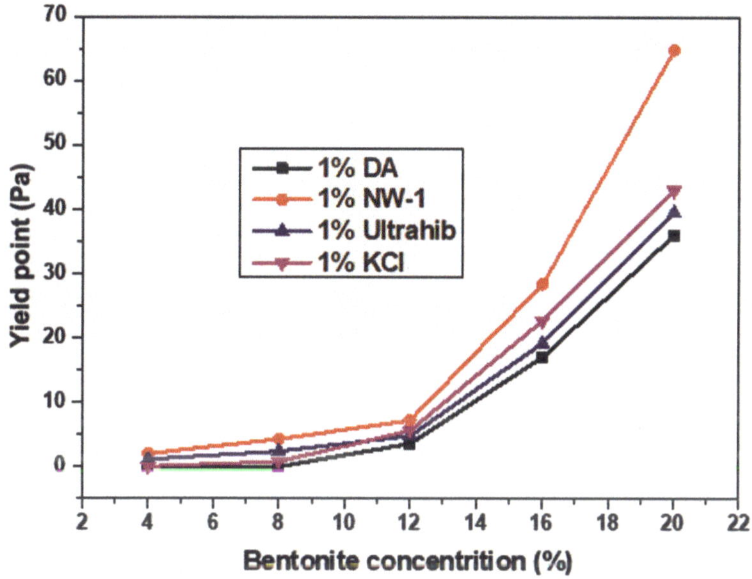

Figure 1.9 Change in YP of different inhibitor/bentonite dispersion systems with amount of bentonite added.

concentrations gradually increases. When there are many bentonite particles, the low concentration of DA molecules is not enough to be close to saturation adsorption on the surface of a large amount of bentonite particles, resulting in extremely limited inhibition effect and large YP. When the DA concentration reaches 1.5%, it can basically completely inhibit the hydration dispersion of bentonite added by 16% and keep the YP unchanged.

Fig. 1.9 compares the inhibition performance of clay mud making with different inhibitors at 1% addition. The results show that at any amount of bentonite, the YP of the DA/bentonite dispersion system is lower than that of several other inhibitor/bentonite systems, indicating that DA has better clay-mud-making inhibition than other commonly used shale inhibitors.

1.2.2.4 Mud ball immersion experiment

The bentonite was mixed with a small amount of water and kneaded into mud balls, and the mud balls were immersed in a 1% solution of different inhibitors for 16 h. The clay hydration inhibition performance of other commonly used inhibitors was compared with that of DA by observing

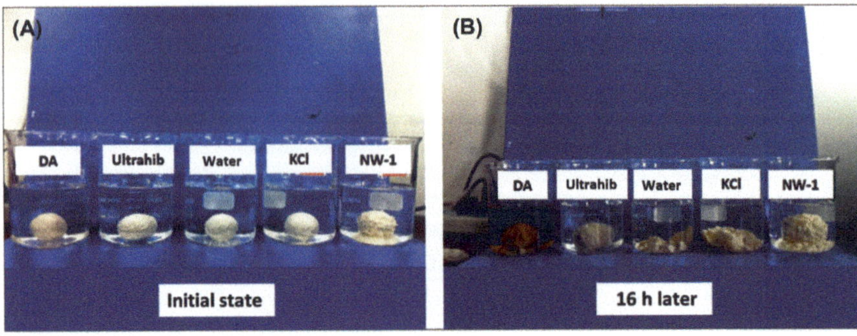

Figure 1.10 Morphological changes of bentonite clay balls after being immersed in different inhibitor solutions with 1% concentration.

the change of the appearance of the mud balls before and after immersion (Fig. 1.10). It can be seen that after being immersed for 16 h, the mud balls that were immersed in pure water and 1% KCl solution completely collapsed owing to hydration expansion, and the mud balls in NW-1 also expanded significantly. The mud balls in DA and polyamine Ultrahib did not collapse nor hydrate significantly. In addition, the color of the DA solution changed from colorless to brown after being immersed for 16 h, which was caused by the spontaneous oxidation of DA in water. In the oxidizing process, DA spontaneously polymerized on the surface of the mud ball, forming a brown pDA membrane, which was also one of the means of protecting the mud ball from collapsing, which will be discussed in detail in the mechanism analysis section.

1.2.2.5 Bentonite interlayer distance experiment

Further, the interlayer distance of bentonite after treatment with different inhibitor solutions was analyzed by X-ray diffraction (XRD) experiment. Based on the diffraction angle corresponding to the peak d001 of bentonite, the interlayer distance d was calculated according to the Bragg equation ($2d\sin\theta = n\lambda$, $\lambda = 1.5406$ nm, $n = 1$) to analyze and compare the ability of different inhibitors to inhibit surface hydration. The experimental results are shown in Table 1.2. The interlayer distance of natural bentonite is 1.29 nm. In contrast, the interlayer distance of samples treated with DA, polyetheramine D230 (MI-SWACO), KCl, and small cations is increased by 0.1, 0.29, 0.17, and 0.24 nm, respectively. The molecular size of DA is about 0.1 nm, indicating that DA molecules were embedded between the clay crystal layers without additional adsorption of water

Table 1.2 Interlayer distance of different sample solutions.

Inhibitor	*d* (nm)
Bentonite	1.29
Bionic inhibitor	1.39
D230 Polyetheramine D230	1.58
KCl	1.46
Small cation	1.53

molecules, which inhibited the occurrence of surface hydration, thereby solving this international technical problem. Other inhibitor products all show the surface hydration phenomena.

1.2.2.6 Effect of pH on the performance of DA

In this section the effect of pH on the inhibition performance of DA was investigated through the experiment of examining the linear expansion and mud-making rate of bentonite in DA solution under different pH conditions.

The linear expansion curves of bentonite in 5% DA solution under different pH conditions are shown in Fig. 1.11. As the pH of the solution increases from 7 to 11, the expansion rate of bentonite increases continuously. In addition, the YP of the DA/bentonite dispersion system under different pH conditions varies with the amount of bentonite, as shown in Fig. 1.12. As the pH value increases from 6.7 to 8.5, the YP of the system at different amount of bentonite increases significantly. These results indicate that the increase of pH value has a negative effect on the clay inhibition performance of DA.

The XRD spectra of 1% DA bentonite adsorbed under different pH conditions are shown in Fig. 1.13. As the pH value increases from 7 to 8.5, the position of the diffraction peak of the bentonite's (001) plane is almost unchanged, and the intensity is weakened. This shows that although the amount of free DA monomer in the system is reduced, it can still polymerize between the layers, generating a double-layer 5,6-dihydroxyindole Indole (DHI) structure pDA. When the pH increases to 11, the position of the diffraction peak obviously shifts to the right to $2\theta = 7.08$ degrees, and the corresponding interlayer distance is reduced to 12.5 nm. When the pH reaches 11, the amount of DA monomer in the system decreases drastically, and it is difficult to adsorb and polymerize a large amount of pDA in the interlayer domain of bentonite to form a

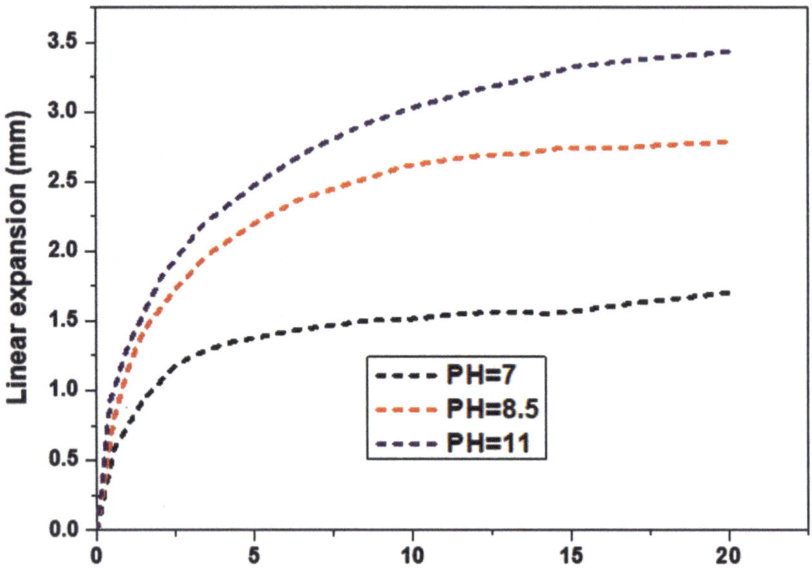

Figure 1.11 Linear expansion curves of bentonite in 5% DA solution under different pH conditions.

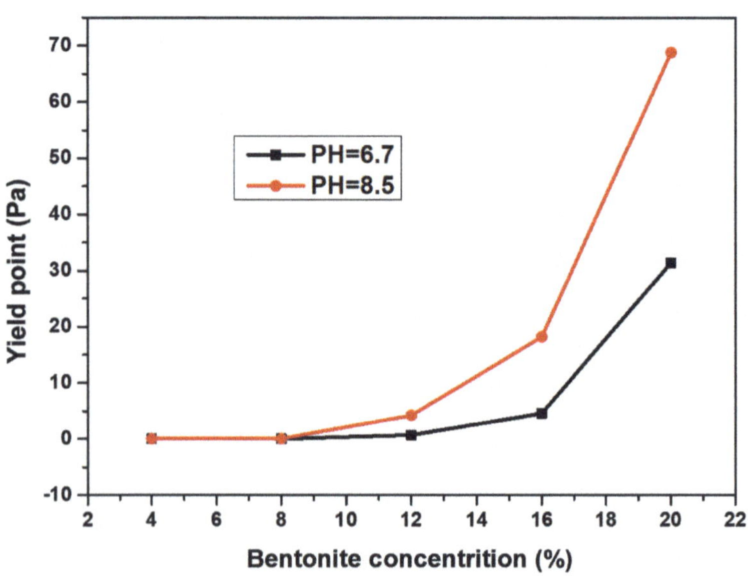

Figure 1.12 Change in YP of DA/bentonite dispersion with amount of bentonite added under different pH conditions.

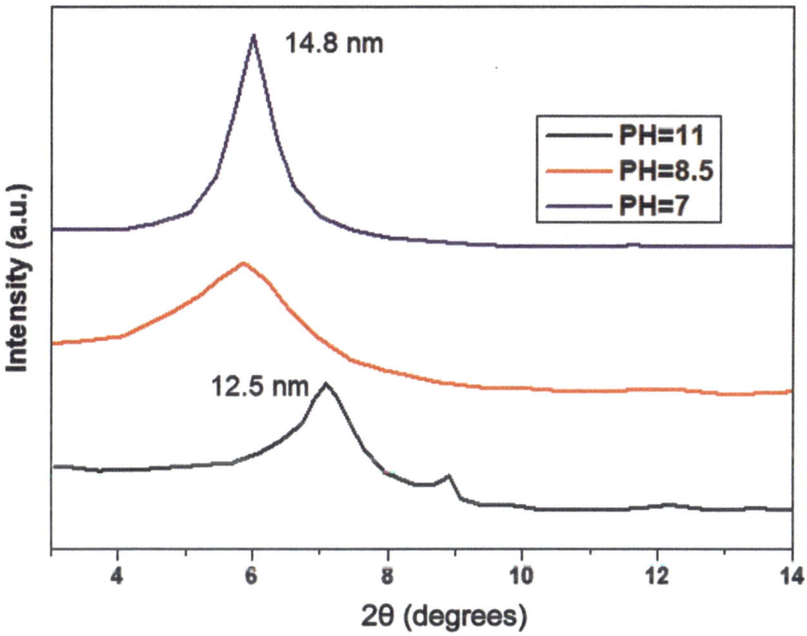

Figure 1.13 XRD spectra of DA-adsorbed bentonite under different pH conditions.

pDA with a more complete structure, which ultimately leads to a great reduction in the inhibitory effect.

On the basis of these analysis results, DA is suitable for drilling fluid systems with a pH value of 7—9; too high a value will weaken the inhibitory effect of DA.

1.2.2.7 Effect of DA on the rheological properties and filtration loss of bentonite base mud

Different amounts of DA were added to different concentrations of bentonite base mud, and the rheological parameters were measured after hot rolling at 120°C for 16 h, from which the influence of DA concentration on the rheological properties of bentonite base mud was investigated. The results are shown in Table 1.3. When the amount of DA added is in the range of 0%—0.5%, it will cause the flocculation of the bentonite particles, resulting in reduced viscosity and increased YP of the system. When the DA concentration is further increased, the bentonite particles will agglomerate, causing the system to increase viscosity, and the filtration loss increases sharply. Therefore to ensure good rheological properties and filtration loss of drilling fluid, the amount of DA should not be too high.

Table 1.3 Effect of DA amount on the rheological properties and filtration loss of bentonite base mud.

Bentonite concentration (%)	DA concentration (%)	Rheological properties and filtration loss					
		AV (mPa · s)	PV (mPa · s)	YP (Pa)	$G_{10''}$ (Pa)	$G_{10'}$ (Pa)	FL_{API} (mL)
2	0	2	2	0	0	0	43
	0.1	2.25	2	0.25	0	0.2	53
	0.2	1.75	1.5	0.25	0	0	44
	0.5	5	0.5	4.5	0.5	1	120
	1	6.25	2.5	3.75	1.25	1.25	310
	1.5	4	2.5	1.5	0.6	0.75	—
4	0	5	4	1	0.1	1.25	29
	0.1	5.75	4.5	1.25	0.25	2.5	33.6
	0.2	5	3.5	1.5	0.75	2.5	37
	0.5	5	2	3	0.75	1.75	40
	1	16.5	4	12.5	5.25	5.5	131
	1.5	9.75	4.5	5.25	3.25	3.25	—
8	0	19.5	14	5.5	2.25	4.25	19
	0.1	18	13	5	7	14	20
	0.2	19.5	9.5	10	8	18	20
	0.5	19.5	10	9.5	7.5	15	24
	1	31.5	7	24.5	14	25	38
	1.5	26	6	20	8.5	8.5	—

1.2.2.8 Evaluation of the temperature resistance of bionic shale inhibitor

The temperature resistance of DA was investigated through thermogravimetric analysis of MMT-DA (powder acquired by drying the bentonite treated with DA solution) with different DA concentrations. The results are shown in Fig. 1.14. The thermogravimetric curves of each sample are generally similar and can be divided into three weightlessness stages: (1) The weight loss in the interval of 30°C–100°C originates from the removal of physically adsorbed water (including interlayer adsorbed water). (2) The weight loss in the interval of 250°C–600°C originates from the thermal decomposition of DA and its derivatives. (3) The weight loss above 1100°C comes from the crystal form transformation of bentonite. Thermogravimetric analysis results show that only when the temperature exceeds 250°C, the inhibition performance of DA or pDA will be greatly weakened, owing to desorption from the bentonite surface.

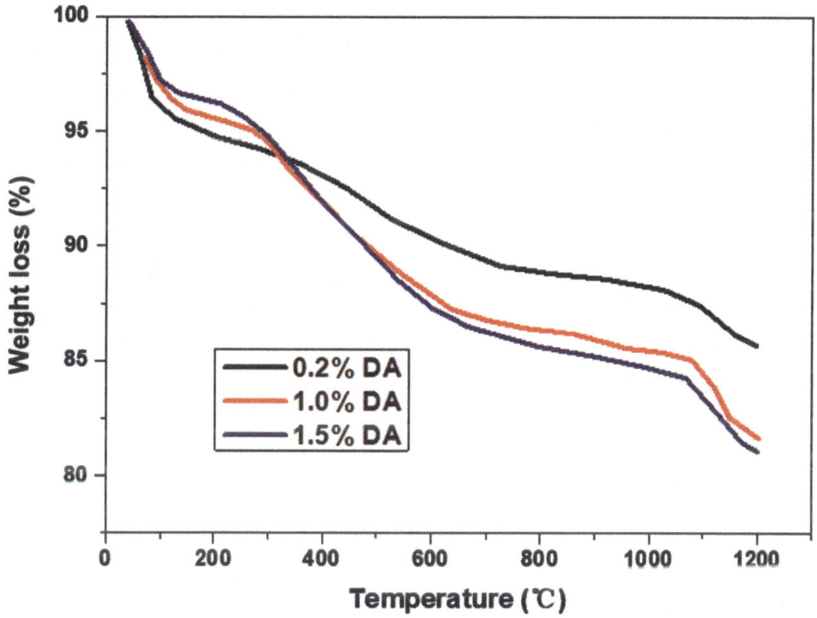

Figure 1.14 Thermogravimetric curves of DA-adsorbed bentonite.

1.2.3 Analysis of the action mechanism of DA

These experimental results show that, in comparison with other inhibitors, DA has superior inhibition performance for shale and clay mud making and solves the problem of surface hydration. This section analyzes the action mechanism of DA by means of infrared spectroscopy, XPS, XRD, transmission electron microscopy (TEM), zeta potential, and Brunauer-Emmett-Teller (BET) pore structure analysis.

1.2.3.1 Mechanism of inhibition for clay hydration

1.2.3.1.1 Infrared spectroscopy analysis

Fig. 1.15 shows the infrared spectra of DA, bentonite, and MMT-DA. The several characteristic peaks of bentonite are the stretching vibration of the structural hydroxyl group ($3625\ cm^{-1}$), the vibration of physical adsorption water ($3443\ cm^{-1}$), the bending vibration of lattice water ($1646\ cm^{-1}$), the stretching vibration of Si–O ($1040\ cm^{-1}$), the bending vibration of Si–O–Si ($1080\ cm^{-1}$), the bending vibration of Al–Al–OH ($917\ cm^{-1}$), the coupling out-of-plane bending vibration of Al–O and Si–O ($625\ cm^{-1}$), and the bending vibration of Al–O–Si

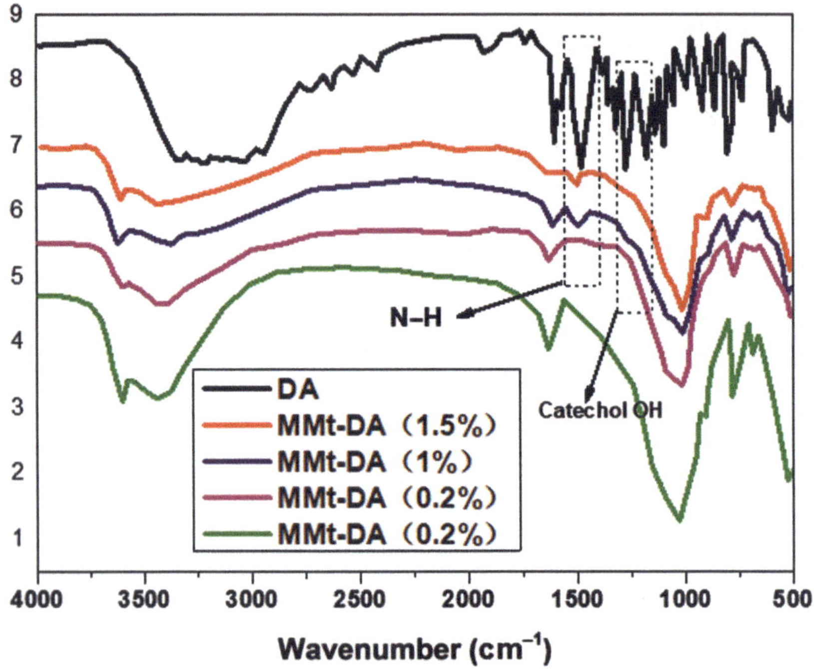

Figure 1.15 Infrared spectra of pure DA and DA-adsorbed bentonite.

(521 cm^{-1}) [13–18]. In the infrared spectrum of MMT-DA, weaker broad peaks appear at 3443 and 1646 cm^{-1}, respectively, indicating that the number of water molecules between the bentonite crystal layers is greatly reduced. In the range of 1200–1600 cm^{-1}, the characteristic peak of DA appears, and the peak intensity increases with the increase of DA concentration, indicating that DA has been adsorbed on the surface of bentonite. A new peak appears at 1447 cm^{-1}, which is due to the polymerization of DA monomer molecules adsorbed between the bentonite crystal layers to form pDA. The weakening of the shoulder peak (bending vibration of Si–O–Si) at 1080 cm^{-1} indicates that the catechol hydroxyl group in DA has hydrogen bonding interaction with the oxygen atom in the bentonite siloxane structure.

1.2.3.1.2 XPS analysis

To further clarify the adsorption behavior of DA on the bentonite surface and the chemical structure of pDA formed by interlayer polymerization, XPS analysis was performed on DA and MMT-DA, with the results shown in Fig. 1.16.

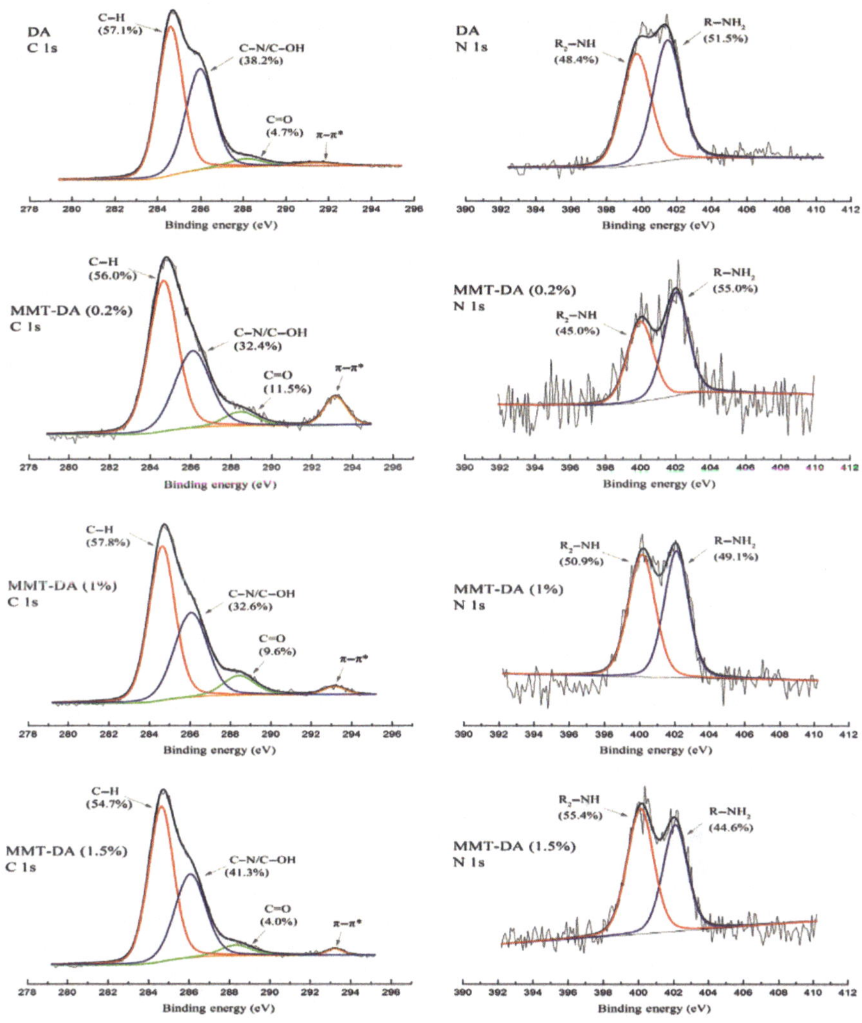

Figure 1.16 XPS results of pure DA and adsorbed-DA bentonite.

In the C1s spectrum of DA, 284.6 eV belongs to the aromatic ring CH, 286.1 and 288.2 eV correspond to C–OH/CN and C=O [19–23], respectively, and the weak peak at 291.4 eV belongs to the transition of aromatic ring π–π*. In the N1s spectrum, 399.7 and 401.5 eV correspond to the two nitrogen-containing functional groups of DA, that is, pyrrole R2–NH and R–NH_2. In the DA molecule the proportion of the functional groups of C=O and R2–NH and C=O is relatively small, while the proportion of R2–NH is relatively large, indicating that

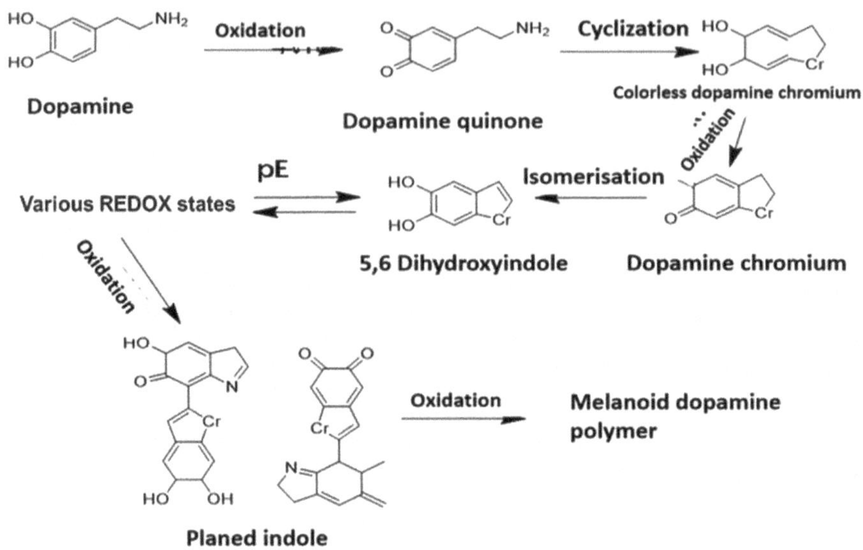

Figure 1.17 Diagram of Raper-Mason mechanism of DA polymerization to form pDA.

DA is metamorphic to leucodopaminechrome owing to oxidation, which is the same as the mechanism of DA oxidation−self-polymerization revealed by Raper-Mason (Fig. 1.17). According to the Raper-Mason mechanism, the polymerization reaction began with DA being oxidized to dopamine quinone and then underwent intermediate steps such as cyclization and isomerization and formed a key intermediate, that is, DHI, and finally DHI monomers continued to polymerize to form melanoid pDA [24−29].

MMT-DA's C1s spectrum differs from DA's in two aspects. One is that the peak of 291.4 eV disappears, and a new peak that does not belong to DA appears at 293.2 eV. According to Subianto's research and the oligomer stacking melanin structure model proposed by Zajac [30−41], this peak is a vibrational companion peak of the $\pi-\pi^*$ transition, derived from the pDA crystal structure formed by the ordered stacking of several layers of DHI oligomers between the bentonite layers. Second, the proportions of aromatic ring C=O and C−OH are different. After DA was polymerized between bentonite layers, the proportions of C=O and C−OH decrease first and then gradually increase with the increase of DA adsorption. This is because DA was oxidized to quinone during the polymerization reaction, and the oxidation degree of interlayer pDA increases as the amount of DA adsorption increases.

In the N1s spectrum the peaks of MMT-DA shift positively by 0.4 eV relative to DA, which is attributed to the protonation of R−NH_2 and pyrrole R_2−NH. When the pH value of the aqueous solution is lower than 8.9, the functional group of R−NH_2 (Pka = 8.9) of DA was easily protonated to R−NH_3^+, and when the adsorption reaction occurred, the DA monomer embedded between the bentonite layers retained the protonated nitrogen-contained functional group. In addition, as the amount of DA adsorption increases, the proportion of functional groups R_2−NH and R−NH_2 in MMT-DA also increases as a result of the closure of the aromatic ring during the polymerization reaction. When the concentration of DA increases to 1.5%, R−NH_2 still occupies a large proportion in MMT-DA. According to Liu's research [42−45], under conditions of controlled pH and deoxygenation, DA monomer can form a monolayer membrane composed of SiO_2 substrate surface without polymerizing into melanoid pDA. Therefore owing to the lack of water and dissolved oxygen between the DA-adsorbed bentonite layers, a large amount of DA monomers cannot be completely oxidized and polymerized. Instead, they are more likely to be distributed on a single layer on the surface of the bentonite crystal layer, which eventually results in the hybrid structure of DHI oligomer sheet and DA monomer between the bentonite layers.

1.2.3.1.3 XRD analysis

Fig. 1.18 shows the XRD patterns of bentonite and MMT-DA. The thickness of the bentonite crystal layer is about 0.96 nm, and the interlayer distance of crystal layers (001 plane) is 1.17 nm, which corresponds to the single-layer hydration state of bentonite [46]. For MMT-DA with a DA content of 0.2% the interlayer distance is increased to 1.28 nm, and the height of the interlayer domain is 0.32 nm, which is slightly smaller than the thickness of the double-layer stacked DHI oligomer (about 0.34 nm). Therefore the pDA between the bentonite layers exists mainly in a single-layer structure. When the DA content is increased to more than 1%, the interlayer distance of bentonite crystal layers further increases to 1.54 nm and does not expand further when the DA content increases, indicating that a complete structure pDA was formed between the bentonite layers. Under this condition, the height of the interlayer domain is about 0.58 nm, which is about the thickness of the double-layer DHI oligomer. This is because when the amount of DA monomer is sufficient to completely cover both sides of the crystal layer, the pDA formed by its polymerization is a double-layer structure. As a result, in the linear

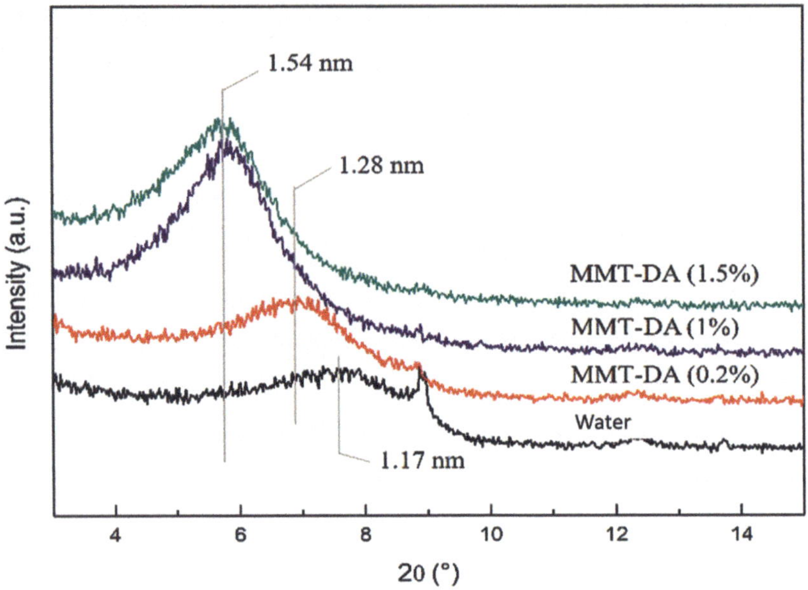

Figure 1.18 XRD patterns of DA-adsorbed bentonite.

expansion experiment, owing to the formation of double-layer pDA, the expansion rate of bentonite in the high-concentration DA solution is higher than that of the low-concentration solution in the initial 2 h. In addition, as the concentration of the DA solution increases, the intensity of the diffraction peak in the MMT-DA spectrum increases and the half-width decreases, indicating an increase in the volume and crystallinity of the bentonite particles. This shows that the high-concentration DA solution is more inhibitory, thereby greatly inhibiting the hydration dispersion of bentonite particles.

1.2.3.1.4 TEM analysis

TEM images (Fig. 1.19) further reveal the mechanism of DA inhibiting the hydration dispersion of clay. The bentonite sheet structure is highly dispersed in water (Fig. 1.19A), but after DA was added (Fig. 1.19B), the bentonite sheet layer stacked and agglomerated to form thicker and larger slab-shaped particles, which intuitively shows DA's good inhibitory effect for clay dispersion.

1.2.3.1.5 Zeta potential analysis

The effect of DA concentration on the hydration dispersion of bentonite was further analyzed by a zeta potential test, with the results shown in

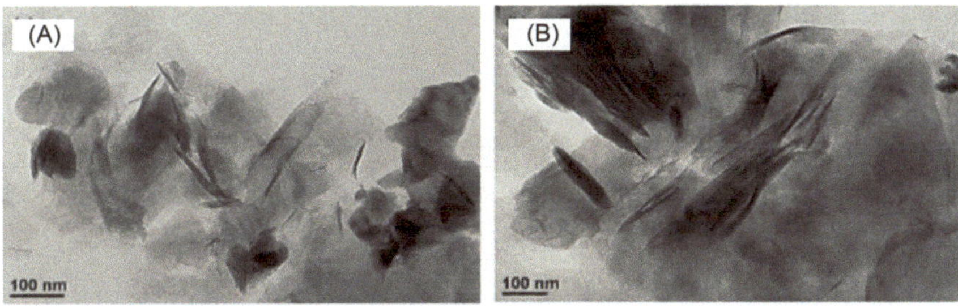

Figure 1.19 TEM images of (A) pure bentonite and (B) DA-adsorbed bentonite.

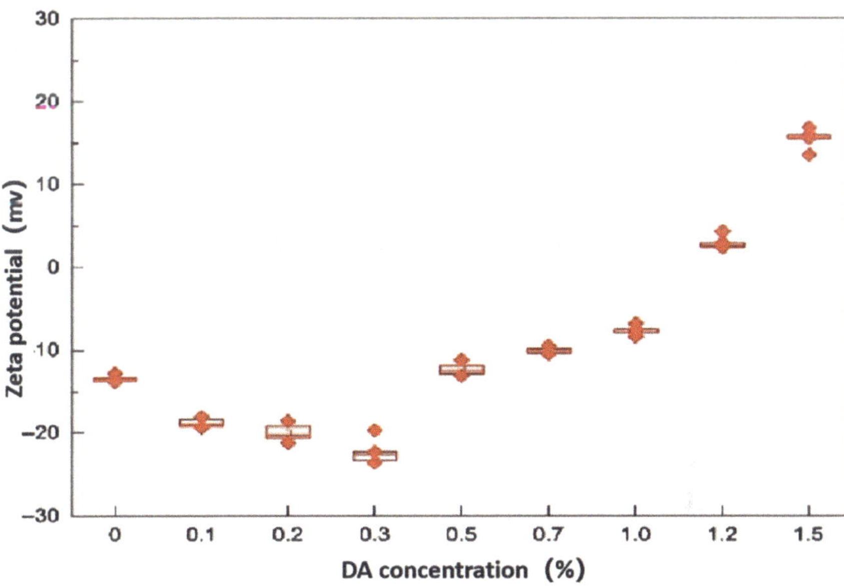

Figure 1.20 Zeta potentials of bentonite dispersed in different concentrations of DA solution.

Fig. 1.20. The lower the zeta potential of the bentonite particles, the better the water dispersibility. The zeta potential of bentonite in water is −13 mV. In low-concentration DA solutions (<0.3%) the zeta potential of bentonite decreases slightly, owing to the negative charge of the pDA polymerized between the layers of bentonite due to the dissociation of quinoneimine (Pka = 6.3). As the concentration of the DA solution continues to increase to 1.5%, the zeta potential value gradually increases and reaches +10 mV, owing to the adsorption of a large number of positively charged DA monomer molecules on the surface of the bentonite, further leading to the collapse of the double layer of the bentonite colloid and effectively inhibiting the hydration dispersion of bentonite.

1.2.3.1.6 Analysis of the gel strength of DA/bentonite dispersion

The change of the gel strength of the DA/bentonite dispersion with DA concentration is shown in Fig. 1.21. With different amounts of bentonite the gel strength curve of each sample shows a trend of increasing first and then decreasing with the increase in DA concentration. This result indicates that with the gradual increase of DA concentration, the bentonite sheet undergoes a process of flocculation to form a 3D network structure (increase in gel strength), followed by a high degree of aggregation due to the collapse of the electric double layer (decrease in gel strength).

The mechanism by which DA inhibits bentonite hydration is illustrated in Fig. 1.22. Owing to ion exchange, electrostatic attraction, and the high adsorption energy of catechol group, a large number of positively charged DA monomer molecules are adsorbed in the interlayer domain of bentonite layer prior to the adsorption of water molecules. This results in the collapse of the double layer of bentonite colloidal diffusion and the occurrence of particle agglomeration. Then DA monomer is polymerized in the interlayer domain of bentonite to form a hydrophobic pDA with a

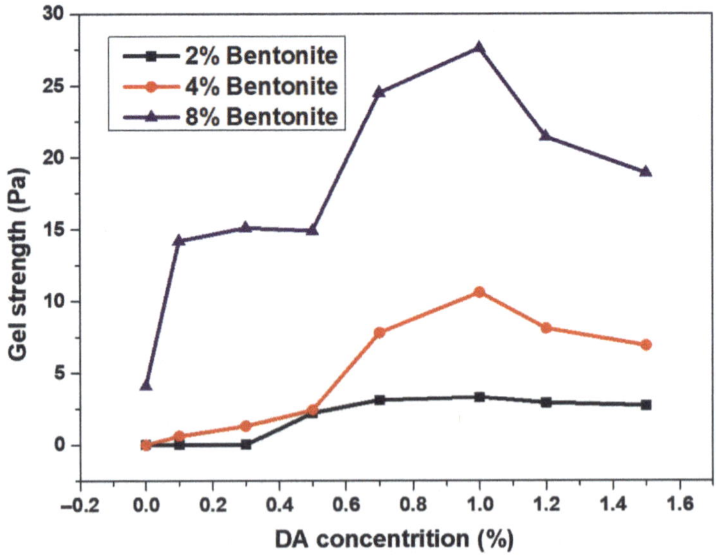

Figure 1.21 Change of the gel strength of DA/bentonite dispersion with DA concentration solution.

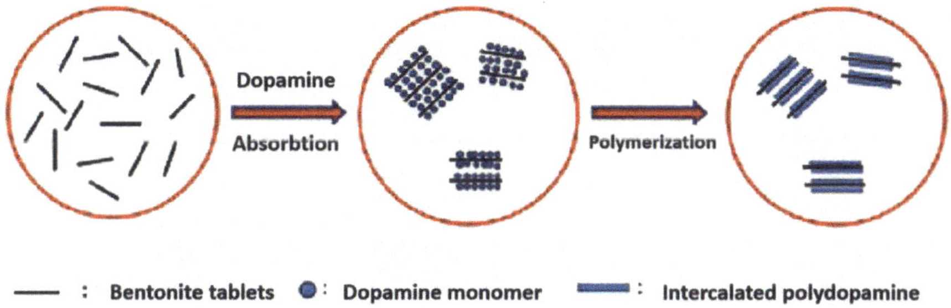

Figure 1.22 Schematic diagram of DA inhibiting the hydration expansion of bentonite.

double-layer DHI structure, and the adjacent bentonite crystals are bonded together by the strong hydrogen bonding between the DHI oligomer and the bentonite siloxane, inhibiting further intrusion of water molecules, thereby inhibiting the occurrence of surface hydration and permeation hydration.

1.2.3.2 Mechanism of membrane formation and plugging of shale

According to the research of Lee et al. [47], DA can spontaneously oxidize and polymerize in an alkaline environment to form a DHI oligomer with a four- to eight-layer structure. These oligomers are further stacked and form a pDA membrane on the surface of an open substrate. Therefore when the drilling fluid containing DA contacts shale, a pDA membrane will also be formed on the pore surface of the shale, thereby partially plugging the micro-nano pores of the shale and thus to a certain extent preventing the intrusion of water into the shale through a capillary force effect. To prove this conclusion, shale cuttings of 6–10 mesh were first rolled in DA solution at 80°C for 16 h and then taken out and dried to examine the changes in microscopic morphology of the shale that formed the pDA membrane by using a scanning electron microscope (SEM). Then the plugging effect of pDA membrane on shale micropores was quantitatively investigated by using the specific surface area and pore size analysis based on BET experiments.

Fig. 1.23 shows the comparison of the microscopic morphology of the shale surface before and after membrane formation. There are a lot of micropores on the surface of the original shale. After the pDA membrane has formed, the surface of the shale becomes very dense, and most of the pores have been plugged by the pDA membrane.

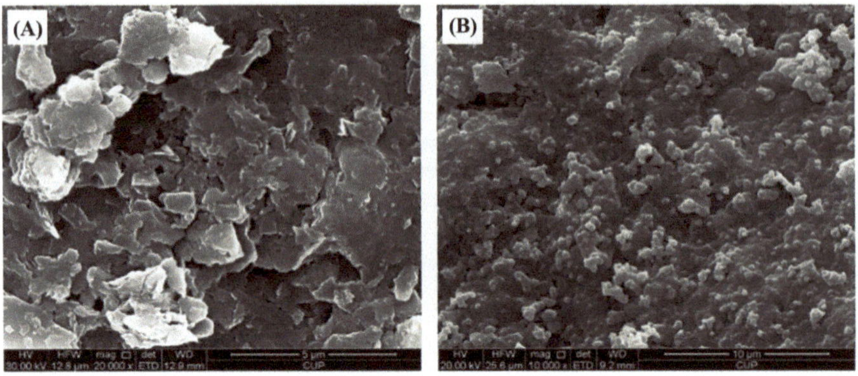

Figure 1.23 SEM images of shale (A) before and (B) after the formation of pDA membrane.

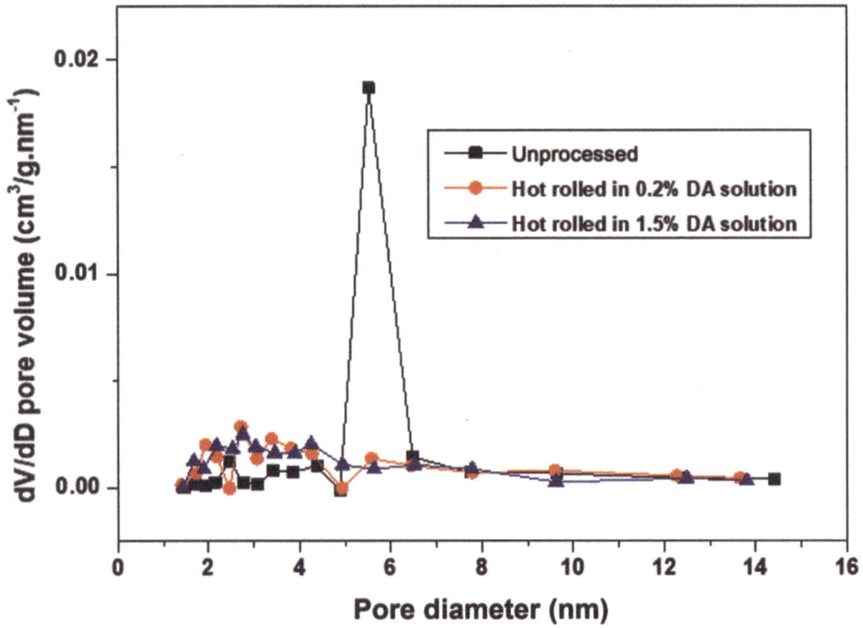

Figure 1.24 Pore size distribution curve of shale after hot rolling in different concentrations of DA solution.

Fig. 1.24 and Table 1.4 show the changes in pore size distribution, specific surface area, pore volume, and average pore size of shale before and after pDA membrane formation. Fig. 1.24 shows that the pore diameter of the original shale is mainly distributed around 6 nm. After hot

Table 1.4 Pore structure changes of shale after hot rolling in DA solutions with different concentrations.

Properties	Shale samples		
	Untreated	0.2% DA	1.5% DA
Pore volume (mL/g)	0.030	0.018	0.019
Specific surface area (m^2/g)	15.29	10.12	10.45
Average pore diameter (nm)	5.56	2.72	2.76

rolling in the DA solution to form the pDA membrane, the pores that were originally distributed around 6 nm basically disappear, and the number of pores in the range of 1–5 nm increases. This indicates that the pDA membrane can reduce the pore diameter of shale. The pore structure analysis results in Table 1.4 further quantitatively demonstrate the plugging effect of pDA membrane on the pores of shale. After the pDA membrane forms, the pore volume and specific surface area of the shale are reduced by about two-thirds, and the average pore diameter is reduced from 5.6 to 2.7 nm, showing DA's excellent membrane formation and shale-plugging effect. When the DA concentration increases from 0.2% to 1.5%, the shale pore structure does not change significantly, indicating that a small amount of DA can be adsorbed on the shale surface and polymerized to form a membrane for effective plugging.

It can be seen that the DA not only can effectively inhibit the hydration expansion of clay, but also can greatly plug the nano-scale pores of shale. The synergistic effect of these two aspects is the reason that DA has excellent shale inhibition.

In summary, the DA has been prepared by imitating the high adhesion characteristics of marine mussels, and its mechanism and performance have been thoroughly investigated and evaluated. The following conclusions are made:

1. Compared with other inhibitors, DA has superior inhibition performance for shale and clay mud making and effectively solves the problem of surface hydration.
2. DA is suitable for drilling fluid systems with a pH of 7–9. To ensure good rheological properties and filtration control of drilling fluid, the amount of DA should not be too high.
3. DA has excellent temperature resistance up to 250°C.
4. DA's inhibition mechanism of clay hydration is as follows: Prior to the adsorption of water molecules, DA can be adsorbed in the bentonite

interlayer domain under various action forces, such as ion exchange, electrostatic attraction, and high adsorption energy of catechol groups. As a result, the double layer of bentonite colloidal diffusion collapses, and particle agglomeration occurs. Then DA molecules form pDA with a double-layer structure, and the bentonite crystal layers adhere together, effectively inhibiting the further invasion of water molecules,

5. In terms of DA's mechanism of shale formation and membrane plugging, DA can form a pDA membrane on the surface of shale, thereby greatly plugging the nano-scale pores of shale.

In short, the introduction of bionics into the development of bionic inhibitors for drilling fluids not only provides better inhibitory effects than current inhibitors (including MI-SWCO's polyamine inhibitors), but also solves the problem of inhibiting surface hydration.

1.3 Development and evaluation of a bionic wellbore strengthener

First, DA was grafted onto PVA to synthesize a bionic wellbore strengthener whose basic properties such as rheological properties and shale inhibition were systematically evaluated. Then a chemical characterization was performed on the gel membrane formed by the bionic wellbore strengthener on the surface of shale, and the mechanism of action of the bionic wellbore strengthener was analyzed. Finally, the core hot rolled or immersed in the bionic wellbore strengthener was subjected to the measurements of uniaxial stress and point load strength to evaluate the strengthening effect for shale wellbore.

1.3.1 Preparation and characterization of the bionic wellbore strengthener

1.3.1.1 Preparation of the bionic wellbore strengthener

First, polyacrylic acid (PAA) was grafted on PVA, then DA was grafted on the resulting PVA-g-PAA branch, and finally polyethylene with PAA and DA bifunctional side chain PVA-g-PAA-g-DA (abbreviated as PVA-PAA-DA) was developed as a bionic wellbore strengthener. The schematic diagram is shown in Fig. 1.25. The reason for choosing PVA−grafted PAA as the main chain of the bionic wellbore strengthener is that polyacrylic polymers can be chelated and cross-linked with Fe^{3+} or Ca^{2+}, Mg^{2+} to form hydrogels [48]. When the PAA polymer comes into contact with the surface of the shale wellbore (especially mudstone) or

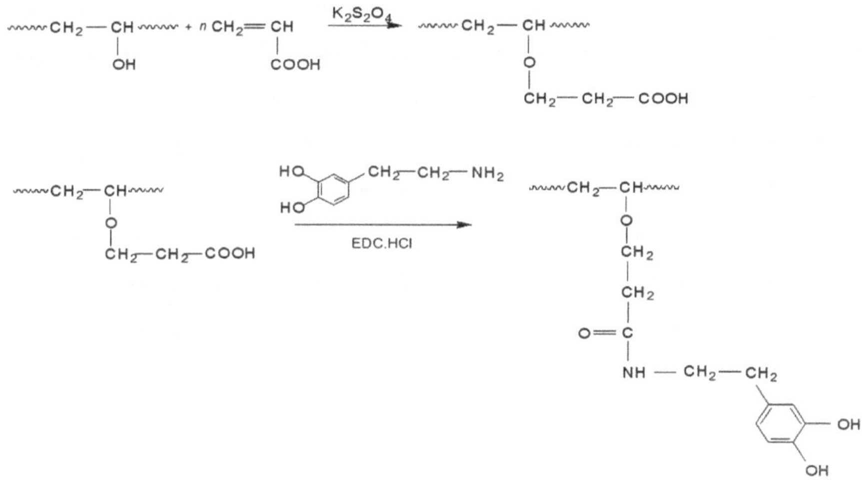

Figure 1.25 Steps of synthesizing the bionic wellbore strengthener PVA-PAA-DA of DA.

penetrates into the pores near the surface, the clay minerals on the rock surface and in the pores will be cemented by PAA to form the hydrogel with certain strength and adhesiveness, thereby strengthening the wellbore by improving the cohesion of the shale. DA can further improve the water-resistant adhesion of the hydrogel and the cross-linking and curing speed on the surface of the shale.

1.3.1.2 Characterization of the bionic wellbore strengthener

1.3.1.2.1 Characterization of PVA-g-PAA

1. Infrared spectroscopy analysis
   Fig. 1.26 shows the infrared spectra of PVA-g-AA and PVA. Both of them show the stretching vibration peak of C=O in the carboxyl group at 1710 cm^{-1}. The difference is that PVA-g-AA shows the stretching vibration peak of C–O–C bond at 1174 cm^{-1}, but the spectrum of PVA does not. Therefore it can be proved that acrylic acid has been successfully grafted on the main chain of PVA.
2. Thermogravimetric analysis
   The thermogravimetric spectrum of PVA-g-PAA is shown in Fig. 1.27. There are three stages of weightlessness. The first stage is 250°C–405°C, a total loss of about 54% of the mass, mainly due to the disconnection of the grafted PAA from the main chain of PVA. The second and third stages are 405°C–510°C and 510°C–1000°C, respectively, mainly due to the pyrolysis of PVA hydroxyl and C–C chain.

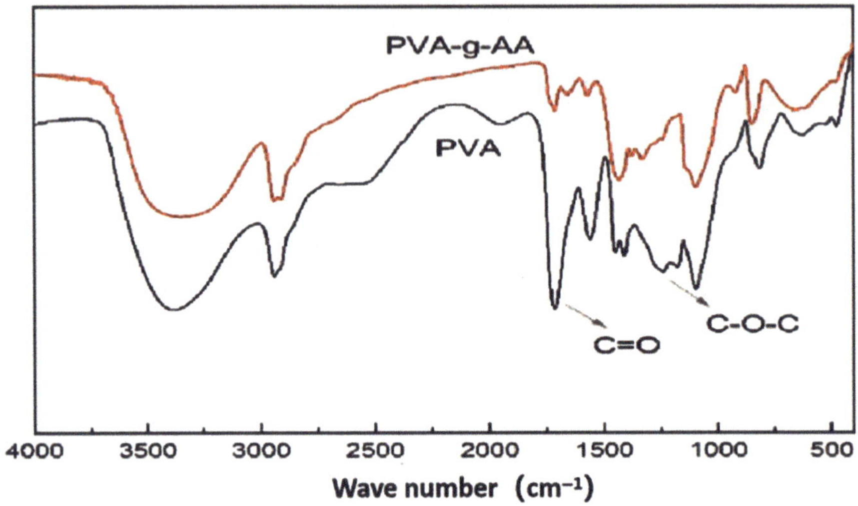

Figure 1.26 Infrared spectra of PVA-g-PAA.

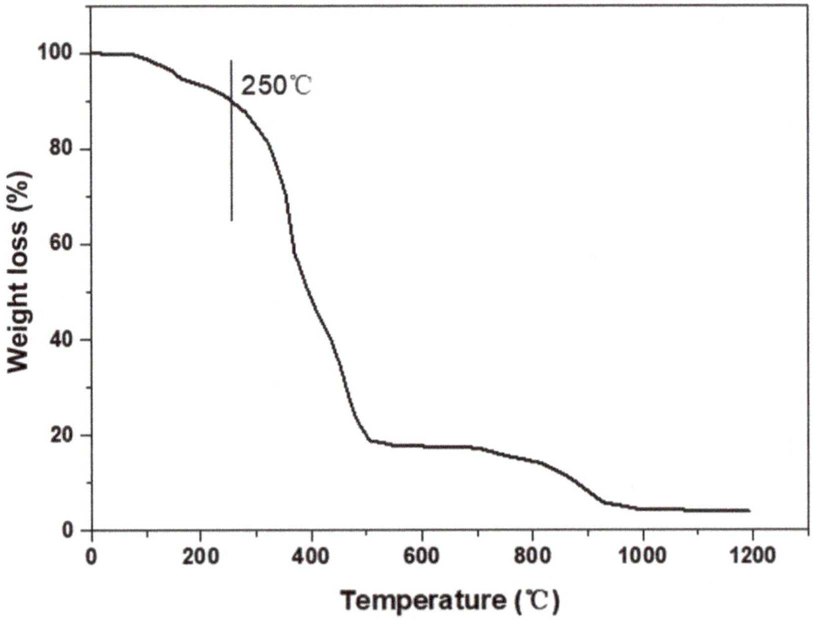

Figure 1.27 Thermogravimetric spectrum of PVA-g-PAA.

1.3.1.2.2 Characterization of PVA-PAA-DA

Owing to the small amount of DA grafted on the bionic wellbore strengthener PVA-PAA-DA, no obvious DA characteristic absorption peak can be found in its infrared spectrum. Therefore XPS analysis was

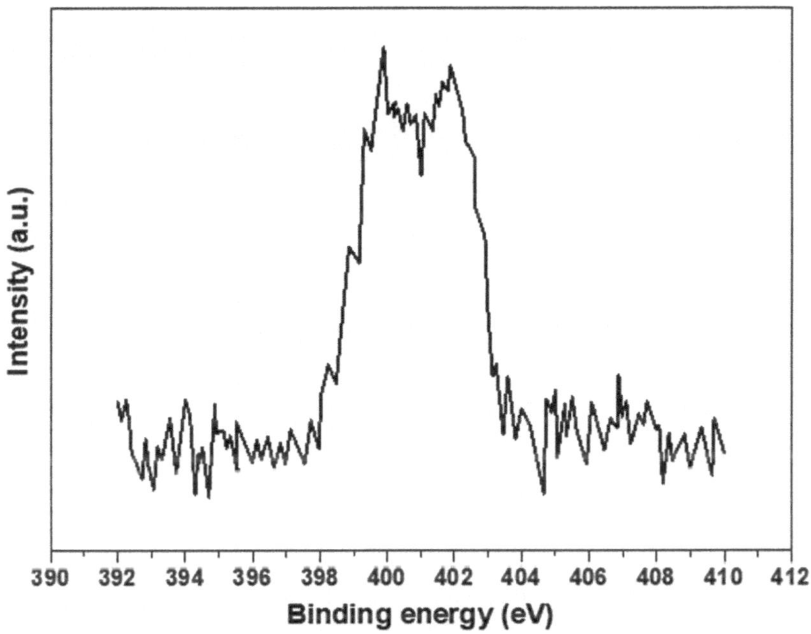

Figure 1.28 X-ray photoelectron spectrum of PVA-PAA-DA (N1s).

performed for characterization, and the results are shown in Fig. 1.28. The peak in the binding energy range of 390–410 eV in the XPS spectrum clearly shows the presence of element N in the sample, indicating that DA has been successfully grafted on PVA-PAA and PVA-PAA-DA has been successfully prepared.

1.3.2 Performance evaluation of the bionic wellbore strengthener

1.3.2.1 Evaluation of the shale-strengthening effect

1.3.2.1.1 Uniaxial stress test of the artificial core

Three pieces of artificial core with the same technical indicators were prepared by compressing quartz sand, bentonite, and cement. Among them, core 1 was not treated, core 2 was rolled in a 0.3% bionic wellbore strengthener solution at 120°C for 16 h, and core 3 was placed in hot water as a control. Because core 3 immersed in water was completely dispersed after hot rolling and the strength test could not be performed, only cores 1 and 2 were subjected to the uniaxial stress test on the rock triaxial servo tester TAW-1000 to evaluate the strengthening effect of the bionic wellbore strengthener on the core. The resulting uniaxial stress-strain curves are presented in Fig. 1.29. It can be seen that the stress of core 1

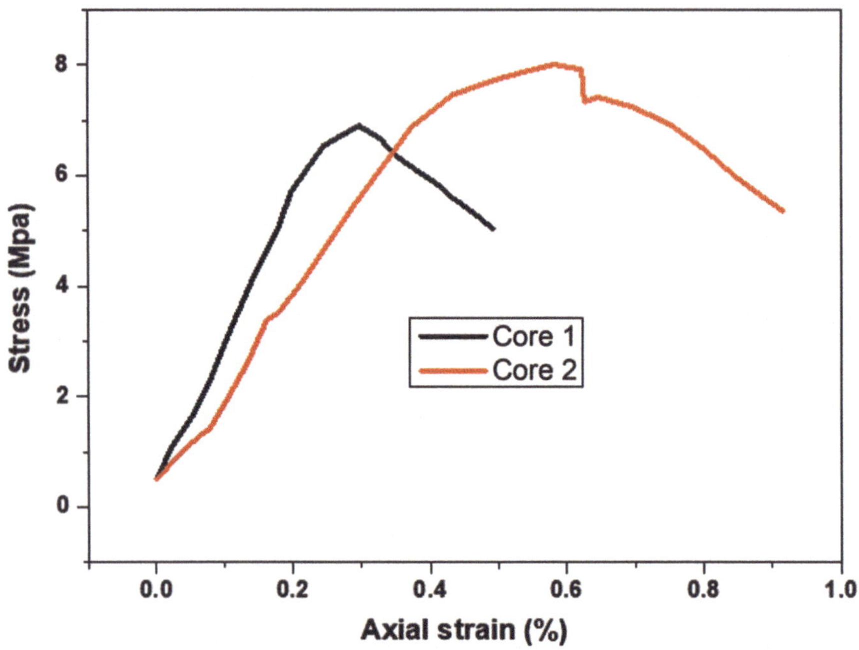

Figure 1.29 Axial stress-strain curves of cores 1 and 2.

Table 1.5 Mechanical parameters of cores 1 and 2.

Core number	Mechanical properties		
	Breaking strength (MPa)	Elastic modulus (GPa)	Poisson's ratio
1	6.906	2.259	0.035
2	8.061	1.855	0.053

reaches its maximum at 6.096 MPa, and the breaking strength of core 2 reaches 8.061 MPa, which is about 15% higher than the original core's strength, showing the good rock-strengthening ability of the bionic wellbore strengthener.

The axial and radial (omitted) stress-strain curves can also be used to calculate the Poisson's ratio μ and the elastic modulus E of the two cores. The results are shown in Table 1.5. The Poisson's ratio of cores 1 and 2 are 0.035 and 0.053, respectively, and their elastic modulus is 2.259 and 1.855 GPa, respectively. Compared with the original core 1, core 2 has a lower modulus of elasticity and a larger Poisson's ratio, indicating that the core after the hot rolling in the bionic wellbore strengthener solution has poor resistance to strain and is soft and plastic. However, the breaking strength of core 2 is significantly higher than that of the original core. This is because the bionic wellbore strengthener invaded the pores of the core, forming a gel with strong adhesion and cohesion, which changed the bonding properties between the core particles.

1.3.2.1.2 Shale point load strength test

To further investigate the shale-strengthening effect of the bionic wellbore strengthener, the artificial shale core pressed from the shale cuttings powder (100 mesh) of Su 53-86-15H horizontal well in the Sulige gas field with high clay mineral content was used for a point load strength test. The total rock mineral composition and clay mineral composition are shown in Table 1.6. The clay mineral content of this core sample is about 33%, of which the proportion of illite is relatively high, reaching 77%, and the content of montmorillonite is 12%.

First, four artificial shale cores were immersed in the solutions of 0.3% bionic wellbore strengthener, 7% KCl, 1% Ultrahib (MI-SWACO high-performance polyamine inhibitor), and 1% polyamine (HPA), respectively, and the changes in the morphology of the core with time were observed. The results show (Fig. 1.30) that only the core in the bionic wellbore strengthener solution remains intact, indicating that its anticollapse ability is stronger than that of several other shale inhibitors. It is worth noting that the clay minerals of the shale used in the experiment are mainly illite, and the content of montmorillonite is relatively low, which is not conducive to KCl's hydration inhibition effect. Moreover, the artificial core compacted in this experiment is loose, and the polyamine inhibitor with lower molecular weight cannot prevent its collapse by an encapsulating effect. It can be deduced from this that when shale that is loosely

Table 1.6 Composition of the tested shale.

Total rock mineral		Clay mineral (%)	
Name	**Content (%)**	**Name**	**Content (%)**
Quartz	29	Kaolinite	2
Potash feldspar	1	Chlorite	10
Plagioclase	12	Illite	30
Calcite	16	Iraqi/Mongolian interlayer	58
Dolomite	6	Interlayer ratio	20
Hematite	3		
Clay minerals	33		

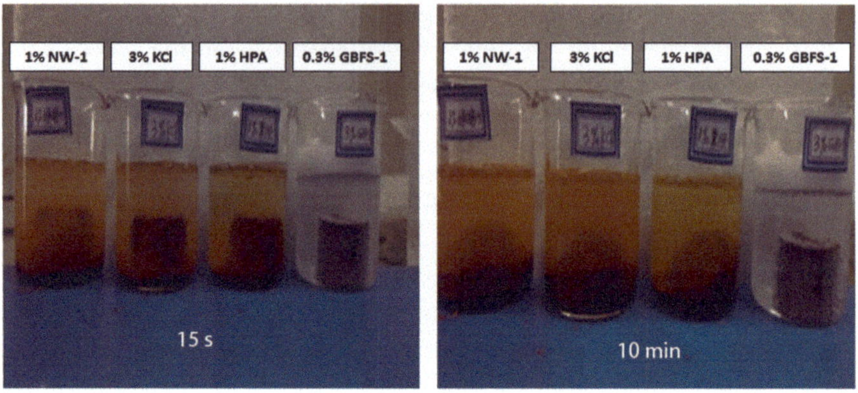

Figure 1.30 Changes in the morphology of shale cores immersed in different solutions.

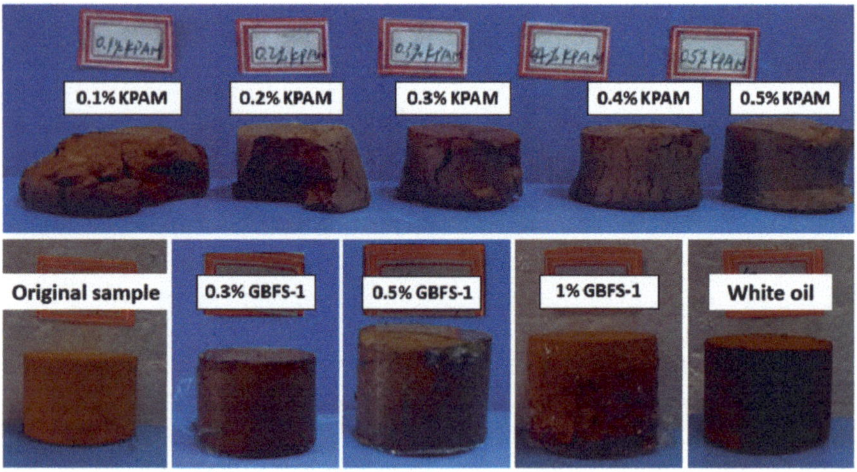

Figure 1.31 Changes in the morphology of shale cores immersed in different solutions (after drying).

cemented or mainly illite is encountered during drilling, the bionic wellbore strengthener has unparalleled advantages in stabilizing the wellbore compared with traditional high-efficiency inhibitors.

Further, the shale cores were immersed in different concentrations of bionic wellbore strengthener solutions and compared with white oil and KPAM solution with higher viscosity and encapsulating effect. The results are shown in Fig. 1.31. It can be seen that although the core immersed in the KPAM solution does not completely collapse, as in other low-molecular weight inhibitors, it has been severely deformed. The core morphology in the bionic wellbore strengthener solution remains basically intact and is only slightly expanded relative to the sample immersed in white oil.

To further quantify the shale-strengthening performance of the bionic wellbore strengthener, the point load strength test was carried out on the original core and the above 10 cores immersed in different liquids (after drying). The load displacement curves were obtained (Fig. 1.32) and were used to calculate the point load strength index of each core, as shown in Table 1.7.

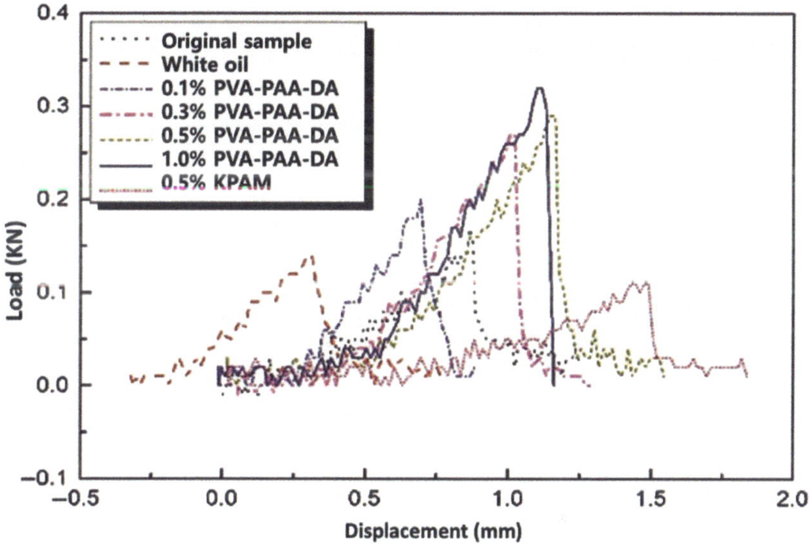

Figure 1.32 Load displacement curves of point load strength test.

Table 1.7 Point load strength of shale cores after being immersed in different liquids.

Solution system	Before-immersion size *h*/*d* (mm)	After-immersion size *h*/*d* (mm)	P_{50} (N)	$I_{s(50)}$ (MPa)
Primitive core	1.90/2.52	1.90/2.52	155	24.41
White oil	1.90/2.52	1.90/2.52	124	15.27
0.1% PVA-PAA-DA	1.90/2.52	1.90/2.66	134	18.94
0.3% PVA-PAA-DA	1.90/2.52	1.90/2.62	196	28.55
0.5% PVA-PAA-DA	1.90/2.52	1.96/2.66	247	34.91
1.0% PVA-PAA-DA	1.90/2.52	1.99/2.63	320	46.26
0.1% KPAM	1.90/2.52	1.64/4.98	49	1.98
0.2% KPAM	1.90/2.52	1.68/3.66	56	4.18
0.3% KPAM	1.90/2.52	1.68/3.10	72	7.49
0.4% KPAM	1.90/2.52	1.76/2.98	81	9.12
0.5% KPAM	1.90/2.52	2.19/2.84	98	12.15

The point load strength of the core immersed in 0.3% bionic wellbore strengthener solution is already higher than that of the original core, reaching 28.55 MPa. As the concentration of bionic wellbore strengthener increases to 1%, the core strength further increases to 46.26 MPa. Although KPAM can also play a certain role in preventing collapse, the strength of the core is only 12.15 MPa when the solution concentration reaches 0.5%. This result further confirms that the bionic wellbore strengthener has better shale-strengthening ability than do traditional anticollapse inhibitors.

1.3.2.2 Evaluation of shale inhibition performance

1.3.2.2.1 Hot rolling recovery experiment

The hot rolling recovery of shale cuttings in different concentrations of bionic wellbore strengthener solutions is shown in Table 1.8. The hot rolling recovery of shale in pure water is 48.9%, and only 0.1% concentration of bionic wellbore strengthener solution can greatly increase the recovery to 78.5%. When the concentration of bionic wellbore strengthener is increased to 1%, the hot rolling recovery of cuttings is further increased to 98.2%. This shows that the bionic wellbore strengthener has an excellent ability to prevent the hydration dispersion of shale.

1.3.2.2.2 Linear expansion experiment

Fig. 1.33 presents the linear expansion curves of shale cuttings in pure water and different concentrations of bionic wellbore strengthener solution. Compared with pure water, the expansion amount of shale in the bionic wellbore strengthener solution is significantly reduced, and it further decreases with increases in the concentration of the bionic wellbore strengthener, indicating that the bionic wellbore strengthener can effectively inhibit the hydration expansion of shale.

Table 1.8 Hot rolling recovery of shale cuttings in the bionic wellbore strengthener solution.

PVA-PAA-DA solution (%)	Initial mass (g)	After-recovery mass (g)	Hot rolling recovery (%)
0	10.0	4.89	48.9
0.1	10.0	7.85	78.5
0.3	10.0	8.97	89.7
0.5	10.0	9.49	94.9
1.0	10.0	9.82	98.2

Note: The hot rolling condition is 120°C and 16 h.

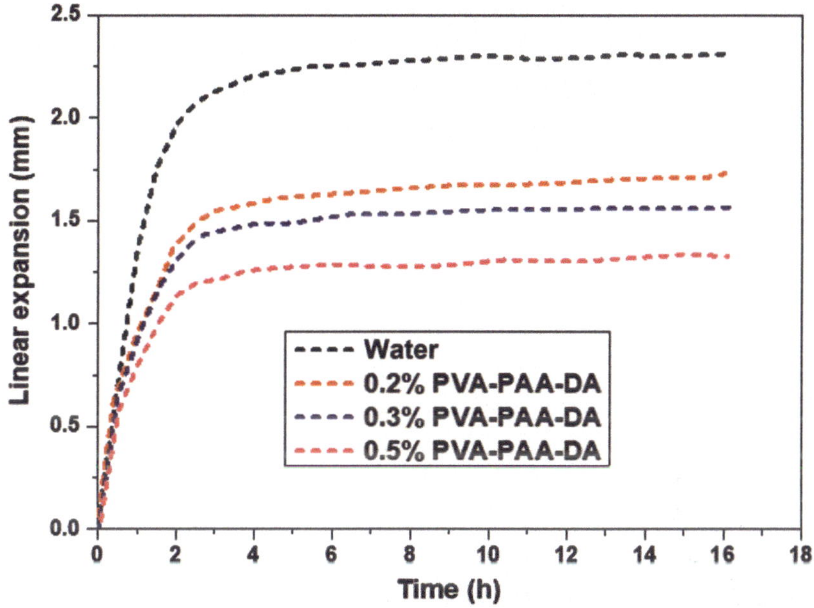

Figure 1.33 Linear expansion curves of shale in pure water and different concentrations of bionic wellbore strengthener solution.

Table 1.9 Point load strength of shale core immersed in bionic wellbore strengthener (concentration of 1%) at different pH values.

pH value of PVA-PAA-DA solution	Before-immersion size *h*/*d* (mm)	After-immersion size *h*/*d* (mm)	P_{50} (N)	$I_{s(50)}$ (MPa)
3	2.00/2.58	2.05/2.63	320	46.26
5	2.00/2.58	2.10/2.60	289	42.75
6	2.00/2.58	2.00/2.62	289	42.10
7	2.00/2.58	2.00/2.60	278	41.12
8	2.00/2.58	2.02/2.60	217	32.10
9	2.00/2.58	2.00/2.62	196	28.55
10	1.90/2.52	2.02/2.62	186	27.10

1.3.2.3 Effect of pH on the performance of the bionic wellbore strengthener

In this section the point load strength of shale cores immersed in bionic wellbore strengthener solutions with different pH values was measured to investigate the effect of pH value on the shale-strengthening performance of the bionic wellbore strengthener. The results are shown in Table 1.9.

As the pH of the solution increases from 3 to 10, the point load strength of the core gradually decreases from 46.26 to 27.10 MPa, indicating that the increase in pH will have a negative impact on the shale-strengthening performance of the bionic wellbore strengthener. This effect is mainly reflected in two aspects. One is that part of the carboxyl group in the bionic wellbore strengthener is neutralized by NaOH in an alkaline environment, thereby changing the configuration and causing its own molecular chain to entangle and curl, resulting in significant reduction in the number of functional groups adsorbed on the surface of shale. The other is that when the pH is greater than 8.5, catechol is accelerated to o-quinone, resulting in weakened interaction with quartz or SiO_2 in clay minerals. However, despite such a negative effect, when the pH is 10, the shale treated with the bionic wellbore strengthener still has higher strength than the original core, showing excellent performance.

1.3.2.4 Effect of the bionic wellbore strengthener on the rheological properties and filtration loss of drilling fluids

This section examines the effect of the bionic wellbore strengthener on the rheological properties and filtration loss of drilling fluids. Table 1.10 shows the rheological parameters and filtration loss of 4% bentonite base mud with different concentrations of bionic wellbore strengthener. As the amount of bionic wellbore strengthener increases from 0.1% to 0.5%, the YP of the dispersion system increases significantly, while the increase in plastic viscosity (PV) is relatively small. This shows that the bionic wellbore strengthener can improve the yield point/plastic viscosity (YP/PV) ratio of drilling fluid to a certain extent and can help to suspend the

Table 1.10 Effect of the amount of bionic wellbore strengthener on the rheological properties and filtration loss of 4% bentonite base mud.

PVA-PAA-DA solution (%)	Rheological parameters and filtration loss						
	AV (mPa · s)	PV (mPa · s)	YP (Pa)	YP/PV (Pa/mPa · s)	$Gel_{10''/10'}$ (Pa/Pa)	FL_{API} (mL)	FL_{HTHP} (mL)
0	9.5	8.5	1	0.12	0.5/1	68	–
0.1	12.25	8.25	4	0.48	0.75/1	18.3	45
0.2	14.5	9.5	5	0.53	1/1.5	17.9	39.6
0.3	16.75	10.5	6.25	0.59	1.75/3	16.4	35
0.5	20.5	14	6.5	0.46	1/1.25	13.6	32

Note: The hot rolling condition is 120°C and 16 h.

weighting materials and clean the wellbore. In addition, the addition of the bionic wellbore strengthener can effectively reduce the filtration loss of the system, probably because the branch of the bionic wellbore strengthener contains a large number of carboxyl groups, and the ability of the molecular chains to form a 3D network structure through hydrogen bonding is relatively strong. Therefore it not only can effectively encapsulate the bentonite particles and prevent the high-temperature coalescence of the bentonite particles, but also can plug the pores of the bentonite cake.

1.3.2.5 Evaluation of the temperature resistance of the bionic wellbore strengthener

First, 0.5% bionic wellbore strengthener was added to 4% bentonite base mud and hot rolled at different temperatures for 16 h. Then the temperature resistance of the bionic wellbore strengthener was investigated through the measurement of rheological properties and filtration loss. The results are shown in Table 1.11. The bionic wellbore strengthener can resist a temperature of 200°C. Although the viscosity will be reduced, owing to the agglomeration of bentonite particles at 200°C (still within the acceptable range), the pressure filtration loss in the system does not change much, with an increase of only 3.2 mL. This shows that the filtration loss reduction mechanism of the bionic wellbore strengthener is mainly the plugging action. The high temperature will not cause decomposition of the polymer chain of the bionic wellbore strengthener, nor will it affect the effective filling of the 3D network structure associated with hydrogen bond in the pores of the cake.

Table 1.11 Effect of temperature on the rheological properties and filtration loss of 4% bentonite base mud containing 0.5% bionic wellbore strengthener.

Temperature (°C)	Rheological parameters and filtration loss					
	AV (mPa · s)	PV (mPa · s)	YP (Pa)	YP/PV (Pa/mPa · s)	$Gel_{10''/10'}$ (Pa/Pa)	FL_{API} (mL)
25	27.5	17	10.5	0.62	1.5/2	10.8
120	20.5	14	6.5	0.46	1/1.25	13.6
150	20	14	6	0.43	0.5/1	11.8
180	4.5	4	0.5	0.13	0.5/0.75	15.0
200	4.25	3.5	0.75	0.21	0.5/0.75	16.2

1.3.2.6 Evaluation of the salt resistance of the bionic wellbore strengthener

First, 0.5% bionic wellbore strengthener was added to 4% bentonite base mud and stirred at 10,000 rpm for 20 min. Then different amounts of NaCl were added and hot rolled at 120°C for 16 h. The NaCl resistance of the bionic wellbore strengthener was investigated through the measurement of rheological properties and filtration loss. The results are shown in Table 1.12. It can be seen that the bionic wellbore strengthener can resist the contamination of 2% NaCl.

1.3.2.7 Evaluation of the calcium resistance of the bionic wellbore strengthener

Similarly, the change in apparent viscosity (AV) of bentonite mud containing 0.3% bionic wellbore strengthener with the addition of $CaCl_2$ (Table 1.13) was evaluated, and the calcium resistance of the bionic wellbore strengthener was evaluated. It can be seen from Table 1.13 that the bionic wellbore strengthener can resist the contamination of 0.5% $CaCl_2$.

Table 1.12 Effect of NaCl on the rheological properties and filtration loss of 4% bentonite base mud containing 0.5% bionic wellbore strengthener.

NaCl solution (%)	Rheological parameters and filtration loss				
	AV (mPa · s)	PV (mPa · s)	YP (Pa)	YP/PV (Pa/mPa · s)	FL_{API} (mL)
0	20.5	14	6.5	0.46	13.6
1	16.5	13	3.5	0.27	21.8
2	15	12	3	0.25	24.2

Note: The hot rolling condition is 120°C and 16 h.

Table 1.13 Effect of $CaCl_2$ on the rheological properties and filtration loss of 4% bentonite base mud containing 0.3% bionic wellbore strengthener.

$CaCl_2$ concentration (%)	Hot rolling situation	AV (mPa · s)	PV (mPa · s)	YP (N/m^2)	YP/PV
0.5	Before hot rolling	16.25	4.5	12.25	2.72
	After hot rolling	19.75	2	17.75	8.875
1	Before hot rolling	14	13	1	0.07
	After hot rolling	17	16	1	0.063

Note: The hot rolling condition is 120°C and 16 h.

1.3.3 Analysis of the action mechanism of the bionic wellbore strengthener

By observing the morphological changes of shale after being immersed in the bionic wellbore strengthener solution, we can get a preliminary understanding of the mechanism of shale strengthening by the bionic wellbore strengthener PVA-PAA-DA. The shale cuttings were immersed in 2% (calculated according to effective component, the same below) bionic wellbore strengthener solution and water, and the morphology of the cuttings was observed after they stood for 1 h. The results are shown in Fig. 1.34.

After being immersed in pure water for 1 h, the originally large pieces of soft shale cuttings were completely hydrated and dispersed to form red mud. However, the cuttings that were immersed in the bionic wellbore strengthener solution did not disperse obviously and were solidified together in a firm way by the gel formed by the bionic wellbore strengthener. At the same time, the color of the bionic wellbore strengthener solution changed from the original dark brown to an almost colorless transparent liquid, which resulted from the bionic wellbore strengthener being precipitated from the solution while being solidified.

The main chain polymer of the bionic wellbore strengthener is PVA with a weight average molecular weight of about 200,000 g/mol. Adding it to the drilling fluid greatly increases the viscosity of the system and

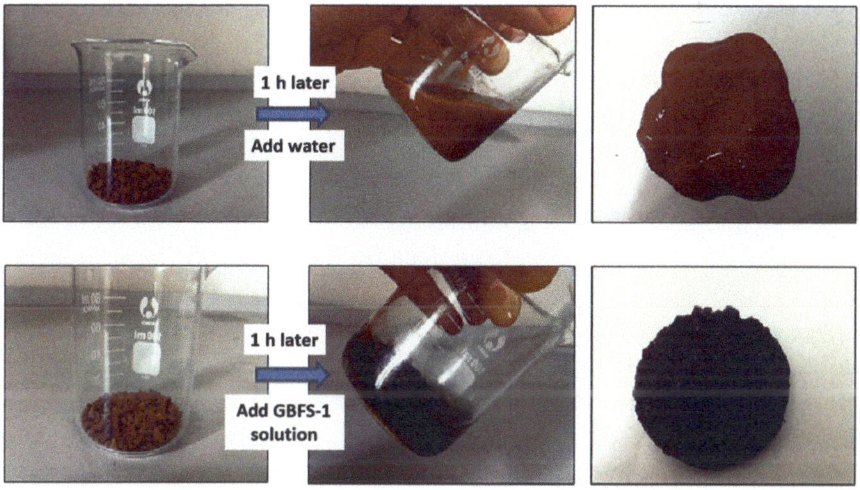

Figure 1.34 Changes in the morphology of shale cuttings immersed in water and 2% bionic wellbore strengthener solution.

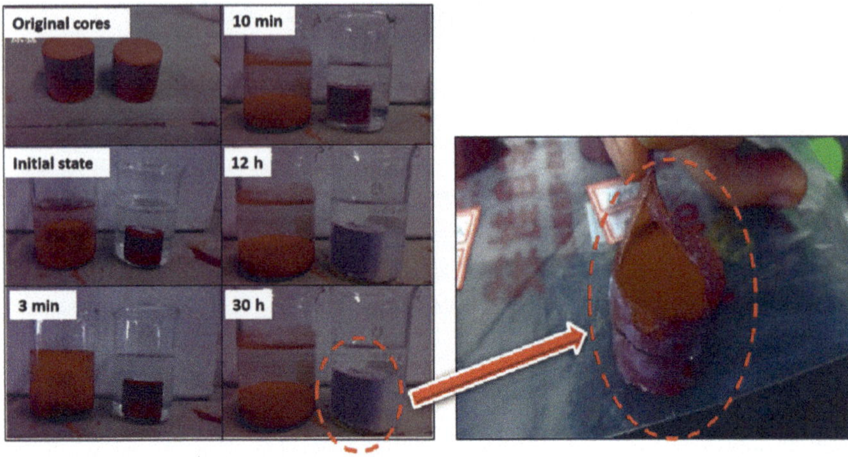

Figure 1.35 Changes in the morphology of shale core immersed in water and 0.5% bionic wellbore strengthener solution.

adversely affects the rheological properties. Experiments have determined that the maximum concentration of bionic wellbore strengthener in the drilling fluid should not be greater than 1%. On this basis the shale-strengthening effect of the low-concentration bionic wellbore strengthener was further investigated. Fig. 1.35 presents the morphology of the shale core after it was immersed in water and 0.5% bionic wellbore strengthener solution. The core that was immersed in pure water completely collapsed after 10 min, and the core that was immersed in the low-concentration bionic wellbore strengthener solution remained intact until 30 h. In addition, after 30 min, a thin gel membrane with relatively low strength formed on the surface of the shale. The results of this experiment show that the low-concentration bionic wellbore strengthener can still play an important role in strengthening the shale.

An elemental analysis was conducted on the shale and the gel membrane formed on the surface of the shale by the bionic wellbore strengthener to analyze the mechanism of spontaneous solidification on the surface of the shale. The experimental results of the elemental analysis are shown in Table 1.14. In addition to the two main elements of Al and Si, the shale contains a small amount of Na, Mg, Ca, Fe, and other elements. The EDS spectra of the gel membrane (Fig. 1.36) show that in addition to the C and O elements, the gel membrane contains a large amount of Ca and a small amount of Mg that are not present in the wellbore strengthener itself. This indicates that the gel membrane was formed by

Table 1.14 X-ray fluorescence spectrum analysis results of the shale.

Element	Content (%)
Al	17.87
Si	55.50
Na	1.32
Mg	3.70
Ca	9.78
Fe	7.02
Others	4.81

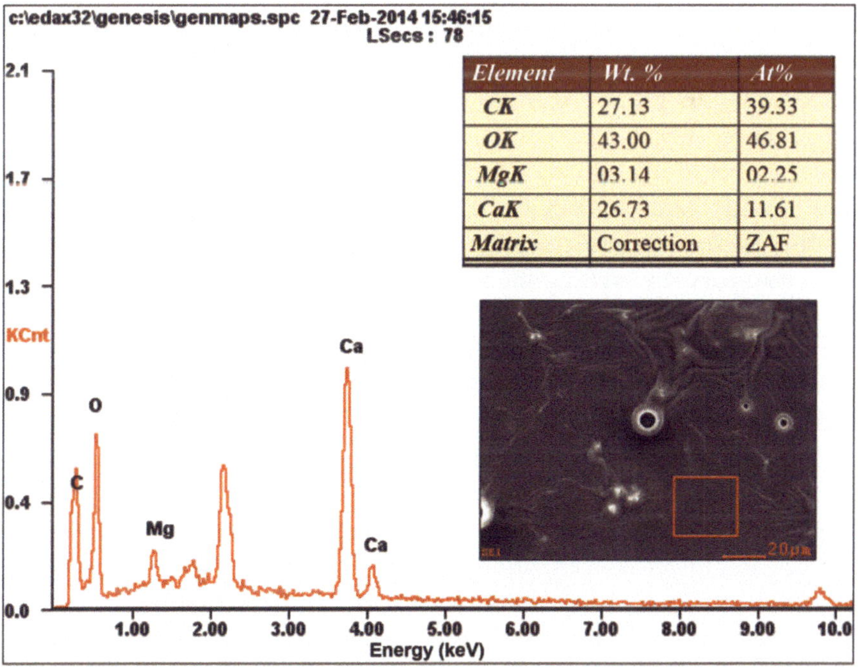

Figure 1.36 EDS spectrum analysis results of the gel membrane formed on the surface of shale by bionic wellbore strengthener.

the chelating and cross-linking reaction between the bionic wellbore strengthener and Ca^{2+} and Mg^{2+} in the shale. In addition, the Ca content in the gel membrane is much higher than the theoretical content in the shale, indicating that during contact of the bionic wellbore strengthener with the shale, Ca^{2+} continuously migrated and concentrated from the interior of the shale to the surface.

Summarizing these experimental results and prior research on the interaction between PAA and clay minerals and the cross-linking of DA-containing polymers with Ca^{2+}, Mg^{2+}, and other metal ions [35,49], we deduce the formation mechanism of the gel membrane on the shale surface by the bionic wellbore strengthener as follows: When the solution of the bionic wellbore strengthener invaded the pores near the surface of the shale, the exchangeable cations (such as Ca^{2+} and Mg^{2+}) inside the shale were continuously replaced by the ionized H^+ of the bionic wellbore strengthener and concentrated on the surface of the shale. Then the macromolecule of the bionic wellbore strengthener adsorbed on the near surface of shale underwent chelate cross-linking reaction with the concentrated Ca^{2+} and Mg^{2+} near the surface through carboxyl and catechol groups, thereby solidifying to form a gel membrane. Owing to the presence of DA groups, this layer of gel membrane has strong cohesion and adhesion, which can significantly improve the bond strength of shale. In addition, the formation of the gel membrane can prevent the further intrusion of water molecules into the shale. These two factors work together to keep the shale wellbore stable.

In summary, by grafting the main components of marine mussel secretions to a special polymer chain, a bionic wellbore strengthener was developed, and comprehensive investigation and evaluation were performed on the mechanism of its action and performance. The following conclusions were made:

1. The bionic wellbore strengthener can solidify on the surface of the shale to form a hydrogel with strong adhesion and cohesion, so it can play a strengthening role in improving the cement strength of the shale at a lower concentration (0.5%).
2. A hot rolling recovery experiment and a linear expansion experiment showed that the bionic wellbore strengthener can greatly inhibit the hydration expansion and dispersion of shale.
3. The bionic wellbore strengthener can improve the YP of the drilling fluid and reduce the filtration loss to a certain extent, which helps to maintain the stability of the wellbore. However, excessively high pH weakens its shale-strengthening performance. It also has certain resistance to temperature and salt. The temperature resistance is at least 150°C, the NaCl contamination resistance can reach 2%, and the $CaCl_2$ contamination resistance can reach 0.5%.

In short, by imitating the principle by which marine mussels' secretion cements rocks, the developed bionic wellbore strengthener does not decrease the rock strength but greatly increases it, which has not been achieved by all the anticollapse agents of drilling fluids in the world, including the most advanced products in the United States. The technical problem of wellbore instability is solved from the internal factors of rock particles, and the purpose of solidification is achieved.

1.4 Development and evaluation of a bionic micro-nano plugging agent

The intrusion and transmission of free water and pressure into the shale formation near the wellbore constitute an important reason for the instability of a wellbore. Therefore effective measures must be taken to prevent the drilling fluid from infiltrating the shale formation. However, owing to the extremely low permeability of shale (10^{-5} to 10^{-11} μm^2), bentonite in the drilling fluid cannot form an external cake on the surface of the shale wellbore [50]. Moreover, traditional drilling fluid plugging materials (such as sulfonated pitch and graphite) cannot penetrate and plug micro-nano pores and microcracks in shale owing to their large particle size [51]. Therefore there is an urgent need to develop nano-scale plugging materials suitable for water base drilling fluids to solve the problem of plugging shale formations that conventional plugging materials cannot achieve and to prevent pressure transmission.

In this section the bacterial structure is used as a model to design the structure of a bionic micro-nano plugging agent, as shown in Fig. 1.37. The bionic micro-nano plugging agent was prepared by using rigid inorganic nanoparticles as the core and grafting flexible polymer. A series of structural characterization and evaluation of plugging performance were carried out. On this basis, the mechanism of action was analyzed. In addition, the effect of the bionic micro-nano plugging agent on drilling fluid performance was investigated.

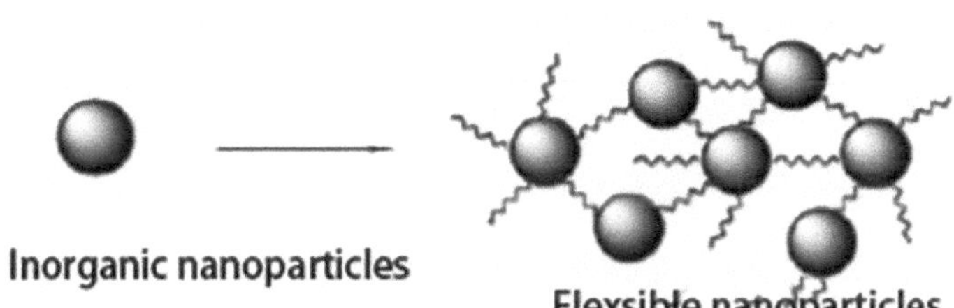

Figure 1.37 Structure design of bionic micro-nano plugging agent (combination of rigidity and flexibility).

1.4.1 Preparation and characterization of the bionic micro-nano plugging agent

1.4.1.1 Preparation of the bionic micro-nano plugging agent

Silica modifier was used to modify the nanosilica. Then the modified nanosilica, pentadienamide-methylene-ethanesulfonic acid, and pentadienamide were used as monomers, and potassium peroxodisulfate was used as the initiator, to synthesize the bionic micro-nano plugging agent CQ-NWD by aqueous solution polymerization.

1.4.1.2 Structural characterization of the bionic micro-nano plugging agent

1.4.1.2.1 Particle size analysis

Fig. 1.38 displays the characterization of CQ-NWD particle size. CQ-NWD has two peaks below 100 nm. In particular, the peak intensity below 10 nm is as high as 48.3%, indicating that it is a nano-scale material, which provides a prerequisite for plugging the shale pores with a size below 10 nm.

1.4.1.2.2 NMR analysis

The H-NMR spectrum of the bionic micro-nano plugging agent is shown in Fig. 1.39. The proportion of hydrogen in the spectrum is

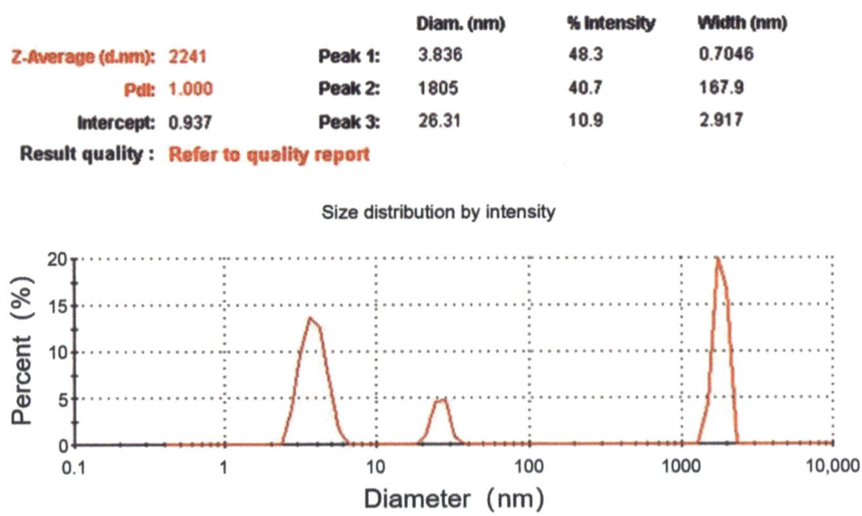

Figure 1.38 Characterization of particle size of bionic micro-nano plugging agent CQ-NWD.

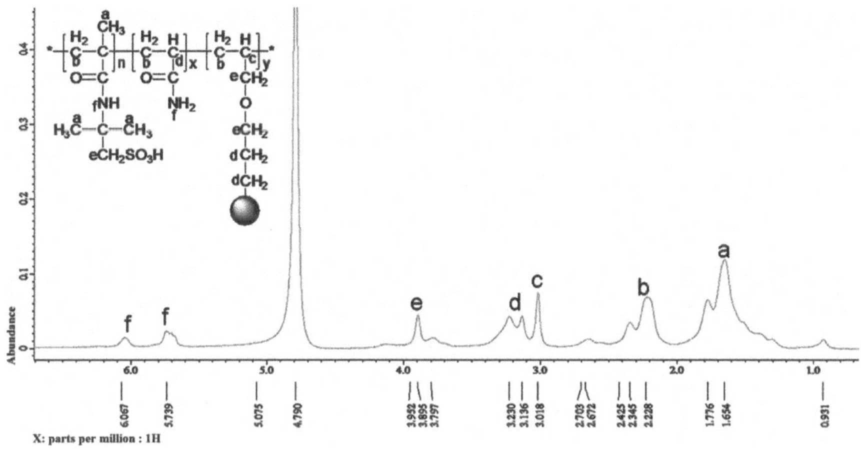

Figure 1.39 ^{1}H-NMR spectrum of bionic micro-nano plugging agent.

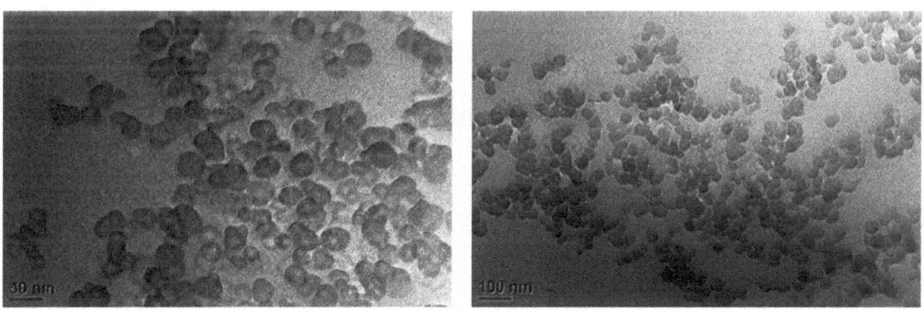

Figure 1.40 TEM images of the bionic micro-nano plugging agent.

consistent with the standards for polymer molecular design, indicating that the products synthesized in the laboratory are of high purity.

1.4.1.2.3 TEM characterization

The TEM characterization of the bionic micro-nano plugging agent is shown in Fig. 1.40. The particles of the synthesized micro-nano plugging agent are mostly below 50 nm. There is partial aggregation between particles, which contributes to forming the aggregates of larger particle size, but such aggregates are easily dispersed under the action of external forces. The experimental results further indicate that the synthesized micro-nano plugging agent is a nanomaterial.

1.4.2 Performance evaluation of the bionic micro-nano plugging agent CQ-NWD

1.4.2.1 Plugging performance evaluation

1.4.2.1.1 Plugging rate test

Fig. 1.41 shows the pore size change before and after plugging the shale by the bionic micro-nano plugging agent. Before CQ-NWD was added, the nano-scale pores of the shale were concentrated at about 6 nm. When CQ-NWD was added, the size of the nano-scale pores was reduced by about 68%, indicating that CQ-NWD has good capacity to plug nano-scale pores. The plugging rate test results by CQ-NWD are shown in Table 1.15. Its plugging rate is as high as 70% or more, which further proves the excellent plugging effect of CQ-NWD.

1.4.2.1.2 Permeability test

The plugging performance of the bionic micro-nano plugging agent was further investigated with a core dynamic simulation device. First, the core was evacuated, then saturated water was used to measure the water

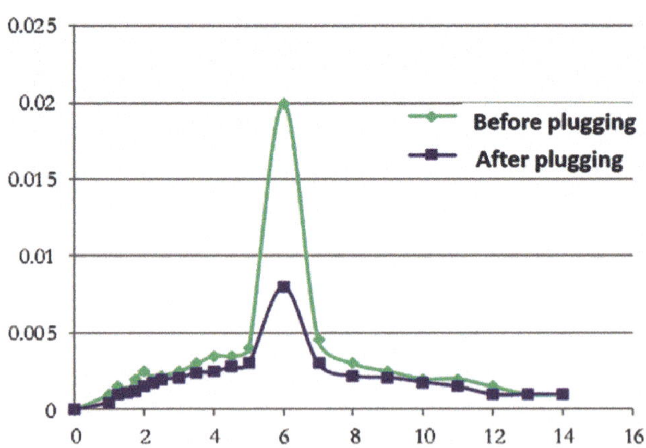

Figure 1.41 Changes in shale pore size before and after plugging.

Table 1.15 Plugging rate test of bionic micro-nano plugging agent.

Sample	60s (mL)	Plugging rate
Pure water	310	—
Sample 1	89	71.3%
Sample 2	84	72.9%
Sample 3	90	71.0%

permeability K0 before plugging, and the CQ-NWD with the same pore volume was injected into a sand-filled tube model with different permeabilities. Two ends of the sand-filled tube were plugged, and the tube was placed at 120°C for 72 h. After that, the solidified sand-filled tube was taken out and reversed to drive the water, and then the water permeability K1 after plugging was measured. The plugging rate of the bionic micro-nano plugging agent was calculated, as shown in Table 1.16. The plugging rate of CQ-NWD is more than 90%, which further proves that the bionic micro-nano plugging agent has an excellent plugging effect.

1.4.2.1.3 Pressure transmission experiment

A shale membrane efficiency test was used to evaluate the plugging performance of CQ NWD. Ultra-low-permeability natural shale core (obtained from Weiyuan, Sichuan) was selected. The confining pressure was 55 MPa, the upstream pressure was 50 MPa, and the change in the downstream pressure was monitored. The results are shown in Fig. 1.42. The downstream pressure in the 4% NaCl brine system began to increase after 5 h. The downstream pressure in the 4% CQ-NWD solution system

Table 1.16 Core permeability test of bionic micro-nano plugging agent.

Sample	K_0 (μm^2)	K_1 (μm^2)	Plugging rate (%)	Breakthrough pressure (mP/cm)
Sample 1	1.923	0.150	92.2	1.96
Sample 2	1.764	0.116	93.4	2.08

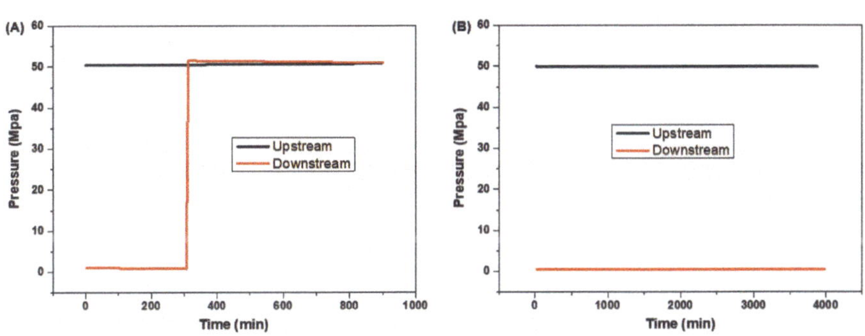

Figure 1.42 Changes of core pore pressure with time. (A) 4% NaCl brine system. (B) 4% CQ-NWD solution system.

Table 1.17 Comparison of the effects of CQ-NWD and other filtration control agents in 4% base mud.

System	AV(mPa · s)	PV(mPa · s)	YP(Pa)	YP/PV	$Gel_{10'/10'}$	FL (mL)
Blank group	7.5	6	1.5	0.25	1/2	30
1% CQ-NWD	19	8	4	0.5	3/5	9.2
1% PAC	32	17	10	0.58	3/4	10.4
1% redul	33	20	13	0.65	3/5	11.6

remained unchanged after 60 h. This shows that CQ-NWD has excellent plugging performance and is more suitable for plugging the pores of ultra-low-permeability shale. It was further proved that the synthesized bionic micro-nano plugging agent particles can enter nano-scale pores.

1.4.2.2 Effect on the rheological properties and filtration loss of drilling fluid

This section evaluates the impact of CQ-NWD on the rheological properties and filtration loss of drilling fluid and compares it with the filtration control agents that are currently used on site. The results are shown in Table 1.17. CQ-NWD has the least effect on the rheological properties of drilling fluid and has the best filtration control effect, highlighting the superiority of nanomaterials.

1.4.2.3 Evaluation of temperature resistance

This section evaluates the temperature resistance of CQ-NWD and compares it with the widely used filtration control agent PAC-LV (Fig. 1.43). Results show that CQ-NWD shows superior temperature resistance.

1.4.2.4 Comparison of salt resistance

The salt resistance of CQ-NWD was evaluated and compared with PAC-LV (Fig. 1.44). The salt resistance of the two is close.

1.4.3 Analysis of the plugging mechanism of bionic micro-nano plugging agent CQ-NWD

According to the one-third plugging theory, the particle size of plugging material should be roughly equal to one-third of pore size. From the characterization experiment results of the bionic micro-nano plugging agent CQ-NWD, it can be seen that the particle size of most of the particles is below 10 nm, which provides a strong foundation for plugging micro-nano pores. However, a comparative test of the influence of

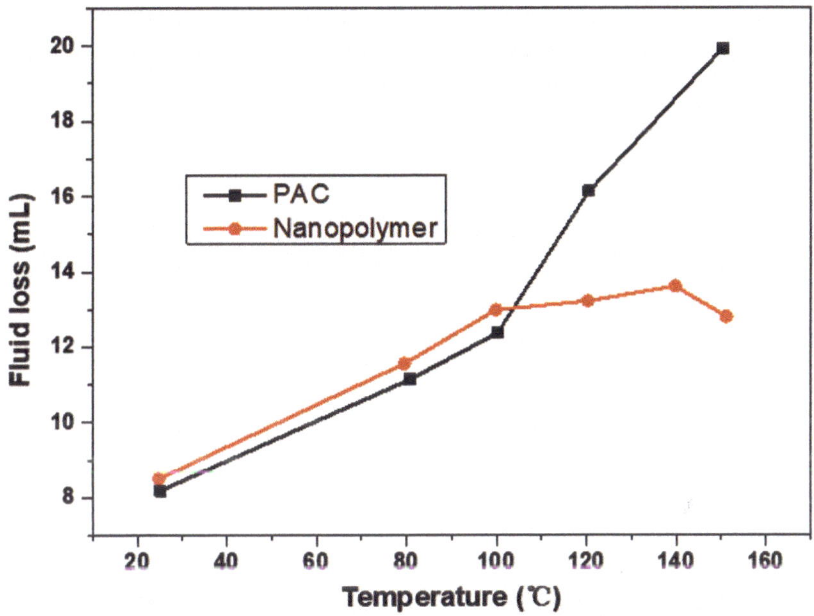

Figure 1.43 Comparison of temperature resistance between the bionic micro-nano plugging agent and PAC-LV.

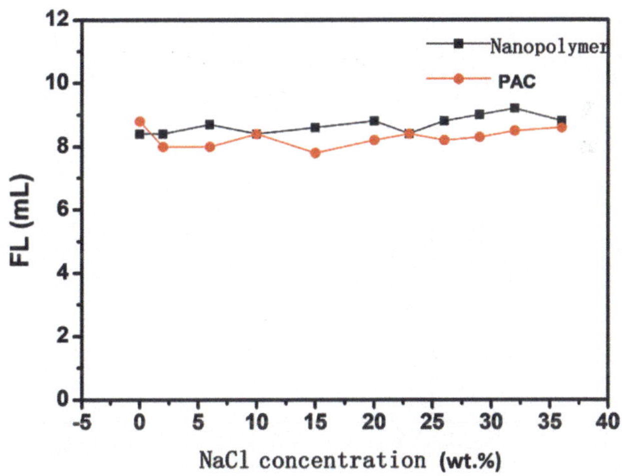

Figure 1.44 Comparison of salt resistance between the bionic micro-nano plugging agent and PAC-LV.

various nano-scale materials on the filtration loss showed that nano size is not the only determinant of plugging effect. In this experiment, to reasonably reflect the plugging performance for nanopores, a hydrophilic polytetrafluoroethylene membrane with a pore size of 220 nm was used instead of ordinary drilling filter paper. The ability of nanosilica, CQ-NWD, and nanosheet materials to plug nanopores is compared in Fig. 1.45. The TEM results of different materials are shown in Fig. 1.46. The particle size of inorganic and simply modified nanosilica is also on the

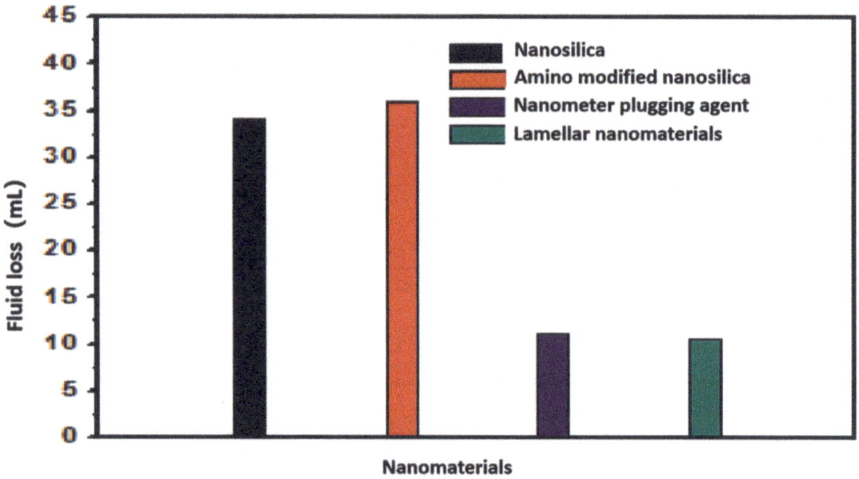

Figure 1.45 Reduced filtration loss of different types of nanomaterials.

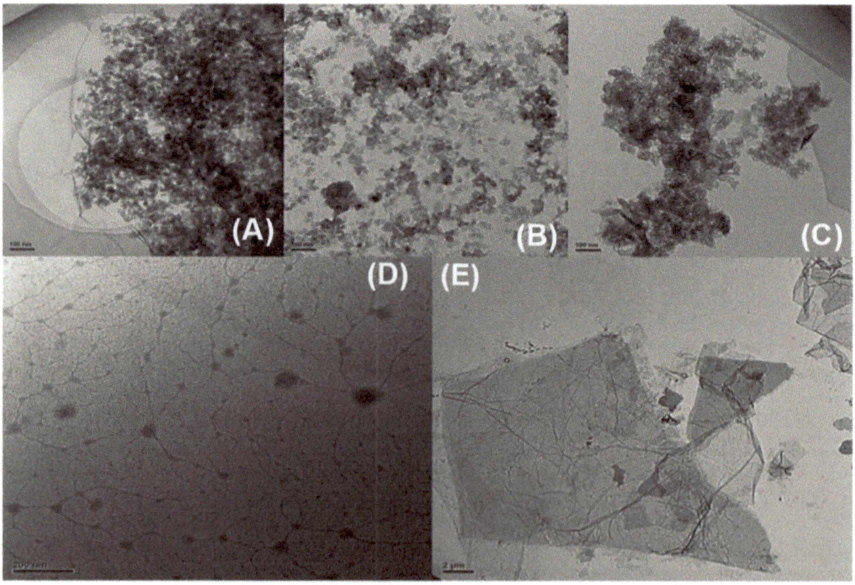

Figure 1.46 TEM images of different nanomaterials. (A) Nanosilica. (B) KH570-modified nanosilica. (C) KH550-modified nanosilica. (D) Micro-nano plugging agent. (E) Sheet nanomaterials.

nano scale, but owing to the lack of functional groups, nanosilica has great limitations in plugging the nanopores of shale, and the amount of filtration is large. The thickness of the lamella nanomaterial is several nanometers, the lamella structure is several micrometers, and its surface is rich in adsorption groups such as hydroxyl and amino groups. The lamellar nanomaterials form a surface membrane by adsorbing on the surface of the medium and plug the nanopores from the outside, so the filtration loss is small, but the ordinary nanomembrane has a relatively low strength and is easily broken under the action of external pressure. CQ-NWD is a low-molecular-weight polymer with a particle size on the nanometer scale and has noncovalent bonds between the molecules to form a spatial grid structure. Adsorption groups such as amide groups on the polymer can be adsorbed to the surface of the external medium. At the same time, low-molecular-weight polymers that do not form a grid structure can also enter the nanopore channels, further plugging the channels. The synergistic effect of the two structures forms a more stable and efficient plugging effect, effectively preventing the passage of water molecules and exhibiting a lower filtration loss.

Therefore the unique action mechanism of bionic micron plugging agent CQ-NWD is that it can spread evenly on the surface of shale with the pressure of the drilling fluid, and the particles in the CQ-NWD connect with each other to form an extremely thin and almost seamless cake, thereby effectively plugging the micro-nano pores and microfractures of the shale (Fig. 1.47). At the same time, the deformable micro-nano particles with strong adsorption groups enter the nano pores of different sizes and are adsorbed on the rock surface at the same time, resulting in a stable plugging effect for nanopores.

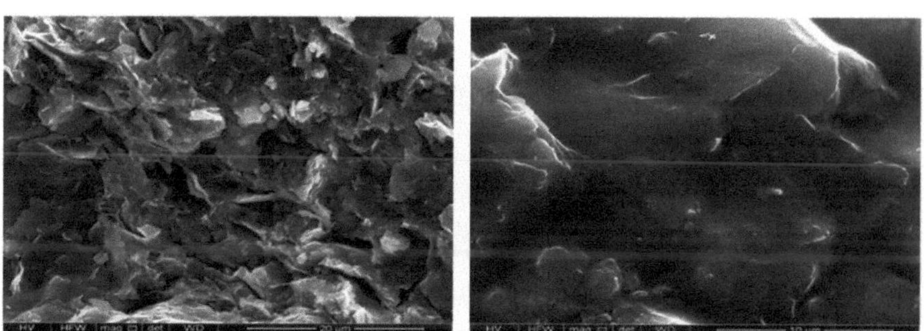

Before plugging **After plugging**

Figure 1.47 Plugging effect of the bionic micro-nano plugging agent.

In summary, by imitating the bacterial structure, the bionic micro-nano plugging agent CQ-NWD was developed. Plugging performance evaluation and mechanism analysis were conducted. The following conclusions were obtained:

1. Particle size characterization and electron microscopy analysis show that CQ-NWD has nanometer size.
2. The results of a conventional filtration control test, a BET pore structure test, and a core dynamic pressure transmission experiment show that CQ-NWD can effectively plug nano-size pores and microcracks.
3. The plugging mechanism of CQ-NWD is that the deformable micro-nanoparticles with strong adsorption groups can be evenly spread on the surface of shale with the drilling fluid pressure and connected with one other to form an extremely thin and almost seamless cake. Meanwhile, some of them can enter different nanopores of different sizes, thus achieving a stable plugging effect for nanopores by the synergistic effect of internal and external factors.
4. Compared with current widely used products, CQ-NWD has superior rheological stability, filtration loss resistance, temperature resistance, and salt resistance.

In short, the developed bionic micro-nano plugging agent has nanoparticle size and nanoparticle characteristics. Furthermore, through the combination of rigidity and flexibility, the micro-nano pores are plugged, and the technical problems that nanoparticles easily form agglomeration and that the plugging material has difficulty plugging the nanopores and cracks are solved, which prevents the pressure transmission of unconventional oil and gas resources, such as coalbed methane, shale oil and gas, and tight oil and gas, and improves the stability of the wellbore.

1.5 New technology and field application of borehole-enhanced bionic water base drilling fluid

The new system of borehole-enhanced bionic water base drilling fluid is formed by taking a bionic wellbore strengthener, a bionic shale inhibitor, and a bionic micro-nano plugging agent as the core agents, supplemented by the necessary viscosifier, filtration control agent, and weighting material. After field test and improvement the new technology of wellbore-enhanced bionic water base drilling fluid with the effects of solidification, inhibition, and plugging is finally obtained. A large number of indoor compatibility experiments have confirmed that the bionic shale inhibitor, bionic wellbore strengthener, and bionic micro-nano plugging agent discussed in this chapter are compatible with almost all conventional drilling fluid agents, so compatibility can be eliminated when screening supporting agents. When using the bionic drilling fluid technology in different oil fields, we can use the drilling fluid systems commonly used in each block or each well team, such as polysulfonate drilling fluid, silicone drilling fluid, and KCl-polymeric alcohol drilling fluid, as the basis and then add appropriate amounts of bionic agents to the chosen drilling fluid to make the system performance meet the requirements.

From 2012 to 2017, wellbore-enhanced bionic drilling fluid systems were successfully applied on site to more than 100 wells in North China Oilfield, Xinjiang Oilfield, Changqing Oilfield, Sichuan Moxi, and other regions, and all have achieved good results. The previously unresolved drilling fluid technical problems have been solved to achieve safe, efficient, low-cost and environmentally friendly drilling. The following is a brief description of the borehole-enhanced bionic water base drilling fluid from the perspectives of preparation method of the field-use formulation, the indoor evaluations of rheological properties, wall building property, wall cementing property and shale inhibition, on-site maintenance, and the effects achieved. The aim is to provide readers with guidance on how to use this new technology and realize safe, efficient, low-cost, and environmentally friendly drilling of different types of complex oil and gas fields under different complex geological conditions.

1.5.1 Field application of bionic drilling fluid in Moxi, Sichuan, China

Wells in the Shaer-Zhenzhuchong formation in Moxi block, Sichuan, China, often suffer from regional collapse. In the drilling process, the suffocation is severe, and the drill hangs easily when being pulled out of hole. Drilling fluids are easily contaminated by soft shale mud making, and thus viscosity increases, which leads to frequent occurrence of downhole complex conditions such as resistance to round trip and bit balling, putting high requirements on the anticollapse inhibition of drilling fluids. From November 2015 to January 2016 a field test of three wells was successfully carried out in No. 119, No. 118, and No. 117 wells in Moxi, using the bionic drilling fluid system. The results show that the bionic drilling fluid has strong anticollapse inhibition performance, the wellbore is stable, the round trip is smooth, and the occurrence of complex downhole accidents is prevented, ensuring the safe and smooth drilling of the upper complex well sections.

1.5.1.1 Bionic drilling fluid system formulation

The upper second-spud drilling fluid was converted into a bionic water base drilling fluid. First, an experiment was conducted indoors, and the formulation was adjusted according to the experimental results. After plugs were drilled, an appropriate amount of Na_2CO_3 was added to remove calcium, the agent was configured into a glue solution according to the formulation, and the solution was then mixed with the upper second-spud drilling fluid to achieve conversion. Before the third spud, the drilling fluid was gradually transformed into the following formulation:

3% Bentonite mud + 0.2% NaOH + 0.2% KPAM + 0.5% PAC-LV
+ 0.8% HGM + 2% SHR-1 + 2% RSTF + 1% MSJ
+ 1.2% bionic inhibitor + 2% bionic wellbore strengthener
+ 1% bionic micro-nano plugging agent
+ 5% KCl + barite.

1.5.1.2 Field application effect

The bionic drilling fluid showed good rheological properties in the on-site test section of Shaximiao to Xujiahe, and the viscosity, YP, and filtration loss were all within reasonable ranges, indicating that the system effectively inhibited shale formation from mud making and enabled the drill of large sections of the shale in this well for the third spud.

Compared with other drilling fluids in adjacent wells, the average borehole diameter expansion rate of the bionic drilling fluid system in different sections was reduced from 12.87% to 5.61%, a decrease of 56.41%, indicating that the bionic drilling fluid showed a strong ability to stabilize the wellbore and prevent the wellbore from collapsing. The results are shown in Tables 1.18 and 1.19.

In the same well section where the drilling fluid density was basically the same, the mechanical penetration rate of the well using the bionic drilling fluid system was greatly increased by 35.4% compared with the No. 10 well using the organic salt polymer drilling fluid system. In addition, the bionic drilling fluid can reduce the time of complex accidents. During the test of the No. 117, No. 118, and No. 119 wells, the average time of redressing by the bionic drilling fluid system was only 4.5 h, and the drilling process was smooth. For the adjacent No. 13 well in the same

Table 1.18 Borehole diameter expansion rate of the bionic drilling fluid in Moxi wells.

Formation	No. 119 well in Moxi			No. 118 well in Moxi			No. 117 well in Moxi		
	Drilling fluid system and density (g/cm^3)	Average borehole diameter (mm)	Expansion rate (%)	Drilling fluid system and density (g/cm^3)	Average borehole diameter (mm)	Expansion rate (%)	Drilling fluid system and density (g/cm^3)	Average borehole diameter (mm)	Expansion rate (%)
Shaximiao	Bionic drilling fluid 1.18–1.50	335.7	4.87	Bionic drilling fluid 1.18–1.47	347.1	5.05	Bionic drilling fluid 1.07–1.53	351.19	6.82
Lianggaoshan	Bionic drilling fluid 1.50–1.57	351.8	5.64	Bionic drilling fluid 1.47–1.61	352.5	5.24	Bionic drilling fluid 1.53–1.71	352.33	5.87
Daanzhai	Bionic drilling fluid 1.54–1.60	330.2	6.11	Bionic drilling fluid 1.61–1.61	355.2	5.68	Bionic drilling fluid 1.71–1.72	353.23	6.17
Zhenzhuchong	Bionic drilling fluid 1.60–1.67	341.8	5.83	Bionic drilling fluid 1.61–1.68	342.3	6.91	Bionic drilling fluid 1.71–1.72	343.57	5.25
Xujiahe	Bionic drilling fluid-high calcium resistance drilling fluid system 1.64–1.70	333.06	4.02	Bionic drilling fluid 1.68–1.75	342.5	5.46	Bionic drilling fluid 1.71–1.75	340.10	5.31

Table 1.19 Borehole diameter expansion rate of other drilling fluid systems in Moxi wells.

Formation	No. 9 well in Moxi			No. 10 well in Moxi			No. 11 well in Moxi		
	Drilling fluid system and density (g/cm^3)	Average borehole diameter (mm)	Expansion rate (%)	Drilling fluid system and density (g/cm^3)	Average borehole diameter (mm)	Expansion rate (%)	Drilling fluid system and density (g/cm^3)	Average borehole diameter (mm)	Expansion rate (%)
Shaximiao	No solid/potassium polysulfonate 1.05–1.53	336.10	8.00	Organic salt polymer 1.07–1.30	336.93	8.29	Organic salt polymer 1.07–1.13	345.20	11.03
Lianggaoshan	Potassium polysulfonate 1.53–1.56	360.60	15.87	Organic salt polymer 1.37–1.40	346.35	11.29	Organic salt polymer 1.12–1.30	375.14	20.55
Daanzhai	Potassium polysulfonate 1.53–1.56	353.71	13.66	Organic salt polymer 1.40–1.50	351.85	13.06	Organic salt polymer 1.24–1.33	371.80	19.47
Zhenzhuchong	Potassium polysulfonate 1.56–1.60	357.53	14.89	Organic salt polymer 1.50–1.51	356.27	14.48	Organic salt polymer 1.32–1.33	355.35	14.19
Xujiahe	Organic salt polysulfonate 1.60–1.65	341.64	9.78	Organic salt polysulfonate 1.51–1.60	339.51	9.10	Organic salt polymer 1.33–1.45	340.63	9.46

Table 1.20 Comparison of the redressing and complex accident handling time of the same well section of the bionic drilling fluid used in Moxi test section and the adjacent well (h).

No. 119 well in Moxi	No. 118 well in Moxi	No. 117 well in Moxi	No. 13 well in Moxi
5.8	0	7.7	65

drilling section of 1190–1542 m (Shayi to Ma'anshan), owing to stratum collapse, three round trips all encountered hang-ups, and the time for dealing with complex downhole time was 65 h. When Yue No. 124 well was drilled to a depth of 1636.91 m (Zhenzhuchong) and pulled out of hole to 1530 m (Ma'anshan), due to strata collapse and getting stuck, the backsliding hole returned 2.5 m^3, and after drilling down to 1600 m (Ma'anshan bottom), the hole was returned to the bottom by 2 m^3. When Yue No. 125 well in 1414–1143 m (Shaer) was being drilled, it took 7 h to start pumping and redressing. drilling down to 1271.34 m, resistance appeared, and it was difficult to make redressing. It took 10.8 h to drill to the bottom. The vibrating screen returned a great number of large cuttings fragments (6–10 mm in diameter) during the drilling. The results are shown in Table 1.20.

1.5.2 Field application of the bionic drilling fluid in Changqing Oilfield, China

The Sulige Gas Field in Changqing Oilfield, China is one of the largest gas fields discovered so far in China. The Su 53 block is a typical low-pressure, low-permeability, low-abundance, strong nonaverage lithologic gas reservoir, and the probability of encountering shale in the horizontal section is high. In the process of shale drilling, complex accidents caused by borehole instability such as diameter reduction, redressing, pipe sticking, collapse block, and resistance will probably occur, and the target layer cannot be reached, which brings great risks to the construction of the horizontal section. To this end, drilling fluid systems such as polysulfonate, silicone, organic salt, and oil base system were used, but the effect was not obvious, which seriously affected the production of Sulige gas field. To solve this long-term technical problem, field tests on the bionic drilling fluid technology with wellbore-strengthening property were conducted from May to October 2013 in three horizontal wells, that is,

Su 53-86-15H1, Su 53-86-15H, and Su 53-70-22H wells in the Sulige gas field. The field test was successful, and the bionic drilling fluid was then widely promoted and applied.

1.5.2.1 Field test formulation of the bionic drilling fluid in Changqing oilfield

The bionic drilling fluid system used in the three test wells of Su 53-86-15H1, Su 53-86-15H, and Su 53-70-22H was obtained based on a silicone formulation. The formulation is as follows:

$$
\begin{aligned}
&2\% - 4\% \text{ Bentonite} + 0.3\% \text{ Na}_2\text{CO}_3 + 2\% - 3\% \text{ organosilicon viscosity reducer}\\
&\quad + 0.6\% - 1.2\% \text{ zwitterionic polymer filtration control agent LX}\\
&\quad + 2\% \text{ FT-1A} + 1.5\% - 2.5\% \text{ bionic wellbore strengthener}\\
&\quad + 0.5\% - 1\% \text{ bionic shale inhibitor} + 1\% - 2\% \text{ liquid lubricant YDRH}\\
&\quad + 2\% - 3\% \text{ ultrafine calcium carbonate}\\
&\quad + \text{barite} \left(\rho = 1.05 - 1.20 \text{ g/cm}^3\right).
\end{aligned}
$$

1.5.2.2 Field application effect

1.5.2.2.1 Su 53-86-15H1 well test situation

A bionic wellbore-strengthening drilling fluid system was used in the Su 53-86-15H1 horizontal wells in the second-spud deviated section and the third-spud horizontal section (3105–4651 m). The strata encountered were mainly in the Shihezi and Majiagou formations. This layer has complex lithology and high shale content, mainly including brown, gray-green shale, light gray and gray-white gravel coarse sandstone, medium-coarse-grained sandstone, gray-green detrital quartz sandstone and gray-brown fine powdered dolomite, deep gray muddy dolomite sandwiched with black shale, dark brown dolomite, and dark gray muddy dolomite. The Su 53-86-15H1 well was drilled from the second spud to completion. The horizontal section encountered the shale at a depth of 384 m, the shale section had a mechanical penetration rate of 4.13 m/h, and the whole well had a mechanical penetration rate of 14.68 m/h. The drilling fluid performance was stable during drilling with the bionic wellbore-strengthening drilling fluid system, and no complicated accident of wellbore instability occurred. It took 30 days to complete the drilling goal. This was the fastest well to be drilled in a single well in Block Su 53. (The horizontal wells in the same block usually required 2 months of drilling under normal circumstances.) The on-site drilling fluid performance is shown in Table 1.21.

Table 1.21 On-site performance of the bionic wellbore-strengthening drilling fluid for the Su 53-86-15H1 well.

Well depth (m)	Density (g/cm^3)	FV (s)	Gel strength (Pa)		Filtration loss (mL)		AV	PV	YP	YP/PV	Coefficient of viscosity	BE (g/L)	*n*	*k*
			10 s	10 min	API	120°C								
3117	1.07	52	2	3	6.5	—	30	23	7	0.31	—	—	0.7	235
3211	1.11	52	2	3.5	5.0	—	32.5	25	7.5	0.3	—	—	0.7	258
3287	1.12	65	2.5	5	4.0	13.6	36.5	27	9.5	0.35	—	42.7	0.66	381
3363	1.13	65	2	4.5	4.0	—	38	28	10	0.36	0.0524	—	0.66	397
3395	1.15	65	2.5	5	3.9	—	43	21	22	1.04	0.0524	—	0.45	983
3497	1.19	65	3	5	3.5	—	44	32	12	0.38	0.0437	—	0.69	492
3568	1.19	65	3	5.5	3.0	—	48.5	35	13.5	0.39	0.0524	—	0.65	546
3748	1.13	64	3.5	11	3.0	12.4	37.5	27	10.5	0.39	0.0524	40.0	0.65	423
3874	1.15	66	2.5	7.5	4.5	—	36	27	9	0.33	0.0437	—	0.68	328
4008	1.15	65	3.5	7	5.0	—	37	26	11	0.42	0.0524	—	0.63	475
4135	1.16	65	2.5	7.5	3.9	—	38.5	29	9.5	0.33	0.0524	—	0.68	351
4291	1.18	61	2	4.5	4.1	—	37	27	10	0.37	0.0524	—	0.64	440
4465	1.19	66	2	5.5	4	—	40	29	11	0.38	0.0524	—	0.65	448
4587	1.20	70	2	6	3.7	—	44.5	32	12.5	0.39	0.0524	—	0.65	501
4651	1.19	60	2	5	3.6	—	41.5	30	11.5	0.38	0.0437	—	0.65	466

Note: *FV*, Funnel viscosity; ρ, density; *BE*, bentonite content; *n*, flow pattern index; *k*, consistency factor.

1.5.2.2.2 Su 53-86-15H well test situation

Su 53-86-15H is the second horizontal well in the same well site as Su 53-86-15H1. The target layer of this well is the fifth and sixth layers of the eighth section of the Shihezi. The lithology of this layer mainly includes brown, gray-green shale, light gray, gray-white gravel coarse sandstone, medium-coarse-grained sandstone, gray-green detrital quartz sandstone and gray-gray brown fine powdered dolomite, dark gray muddy dolomite black shale, dark brown dolomite, dark gray muddy dolomite. Starting from the second spud to the completion of the Su 53-86-15H well, the shale section was up to 1028 m (644 m continuous shale in deviated section, 384 m continuous shale in horizontal section). The performance of the bionic drilling fluid with wellbore-strengthening property showed good stability during drilling and greatly reduced the occurrence of complex accidents such as wellbore instability.

The designed well depth of Su 53-86-15H is 4824.69 m, and the horizontal section is 1200 m long. The shale content of the deviated section is high, and all the well sections from 2963 to 3607 m are shale layers. However, owing to the strong inhibitory bionic wellbore-strengthening drilling fluid system, the total drilling time of the deviated section of this well is only 48 h, and the penetration rate is 13.42 m/h, which is the highest among the deviated sections of all horizontal wells in Su 53 block. During drilling in the horizontal section up to 4250 m, a large section of shale was encountered. During continuous drilling of the 384-m shale section, the drilling time was 112.5 h and the mechanical penetration rate was 3.41 m/h. The average penetration rate of the whole well was 9.45 m/h. The on-site drilling fluid performance is shown in Table 1.22.

The test results of the bionic wellbore-strengthening drilling fluid system in the Su 53-86-15H well show that the drilling fluid has stable performance, good rheological and wall-building properties, excellent cuttings-carrying effect, low viscosity coefficient of cake, and simple performance maintenance. In drilling long sections of shale, there were no downhole complex conditions related to drilling fluids. Compared with the average data of 11 horizontal wells using other drilling fluid systems in the same block (Table 1.23), this well has the longest encounter with shale (644 m continuous shale in the deviated section and 384 m continuous shale in the horizontal section). The average ROP is the fastest (9.45 m/h), and the drilling fluid cost is the lowest (reduced by 36%), which has created the record of the safest downhole conditions and the fastest penetration rate among all drilling fluids used in shale section at that time.

Table 1.22 On-site performance of the bionic wellbore-strengthening drilling fluid for the Su 53-86-15H well.

Well depth (m)	Density (g/cm³)	FV (s)	Gel strength (Pa)		Filtration loss (mL)		AV	PV	YP	YP/PV	*n*	*k*
			10 s	10 min	API	120°C						
3600	1.06	56	2.0	3.5	4.5	14.0	34	24	10	0.42	0.7	235
3700	1.09	58	2.5	3.5	4.5	–	36	25	11	0.44	0.7	258
3800	1.12	64	2.5	5.0	4	13.8	39	28	11	0.39	0.66	381
3900	1.13	66	2.5	4.5	4	–	37.5	27	10.5	0.39	0.66	397
4000	1.16	69	2.5	5.0	3.7	–	39	28	11	0.39	0.45	983
4100	1.19	67	3.0	5.5	3.5	–	41	29	12	0.41	0.69	492
4200	1.19	64	3.0	5.5	3.2	11.8	36.5	27	9.5	0.35	0.65	546
4300	1.19	62	3.5	11	3.6	–	34	24	10	0.42	0.65	423
4400	1.19	63	2.5	7.5	3.5	–	37	26	11	0.42	0.68	328
4500	1.18	63	3.5	7.0	3.9	–	40.5	27	13.5	0.50	0.63	475
4600	1.19	62	3.0	7.0	3.9	–	37.5	25	12.5	0.50	0.68	351

1.5.2.2.3 Su 53-70-22H well test situation

The bionic wellbore-strengthening drilling fluid was used for drilling the third-spud horizontal section of Su 53-70-22H well. The horizontal section (3567–4767 m) is mainly located in the Shihezi and Majiagou. The lithology of this layer mainly includes brown, gray-green shale, light gray, gray-white gravel coarse sandstone, medium-coarse-grained sandstone, gray-green detrital quartz sandstone and gray–gray brown fine powdered dolomite, dark gray muddy dolomite black shale, dark brown dolomite, dark gray muddy dolomite. From starting to completion the horizontal section encountered a large section of shale four times. The total length of the shale section is 410 m. During drilling, the drilling fluid performance remained stable, and the test was successful. The on-site drilling fluid performance is shown in Table 1.24.

The drilling practice of Su 53-70-22H well shows that, after adopting the borehole-enhanced bionic drilling fluid technology, the average ROP increased by 27% compared with other drilling fluid technologies in adjacent wells, and the drilling fluid cost savings was more than 30%, which is very satisfactory.

Table 1.23 The situation of 11 Su 53 wells using other drilling fluid systems.

No.	Well number	System	Total material cost (CNY)	Unit cost (CNY/m)	Type of accident	Lost time (h)	Cost (CNY)
1	53-74-41H	Polysulfonate system	812,255	171.5	Redressing	79	33,479.3
2	53-74-40H	Silicone system	736,840	170.2	Redressing	84	18,538.8
3	53-70-28H	Silicone system	1,662,545	347.7	Lost circulation	183	275,980
4	53-74-42H	Silicone system	834,158	177.6	Pipe sticking	10	Not counted
5	53-78-12H	Polysulfonate system	161,4009	343.9	Redressing/lost circulation	Not counted	Not counted
6	53-78-24H1	Silicone system	1,012,918	218.9	Tool drop	Not counted	Not counted
7	53-78-26H1	Polysulfonate system	743,309	159.8	Not counted	Not counted	Not counted
8	53-78-28H1	Polysulfonate system	855,199	185.2	Lost circulation	73	136,174
9	53-82-14H1	Polysulfonate system	1,130,450	241.5	Lost circulation, well blowout, pipe sticking, redressing	Not counted	Not counted
10	53-82-41H1	Silicone system	523,246	115.5	Not counted	Not counted	Not counted
11	53-86-17H1	Silicone system	802,972	171.1	Lost circulation/pipe sticking	Not counted	Not counted

Table 1.24 On-site performance of bionic wellbore-strengthening drilling fluid for the Su 53-70-22H well.

Well depth (m)	Density (g/cm^3)	Viscosity(s)	Gel strength (Pa)		Filtration loss (mL)		AV	PV	YP	YP/PV	*n*	*k*
			10 s	10 min	API	120°C						
3600	1.13	53	1.5	3.5	5.4	–	28	20	8	0.40	0.65	235
3700	1.14	54	2.0	3.5	5.2	–	18.5	14	4.5	0.32	0.68	258
3800	1.14	53	2.0	4.5	4.6	–	37.5	28	9.5	0.34	0.66	381
3900	1.14	56	2.5	4.5	4.5		37	27	10	0.37	0.65	397
						12.8						
4000	1.14	55	2.5	5.0	4.5	–	35	26	9	0.38	0.65	983
4100	1.14	53	2.5	5.0	4.5	–	30	22	8	0.36	0.67	492
4200	1.14	53	3.0	5.5	4.5	–	23	16	7	0.44	0.65	546
4300	1.14	52	3.0	8.0	4.0	–	35.5	27	8.5	0.32	0.65	423
4400	1.15	56	2.5	7.5	4.0		23.5	17	6.5	0.38	0.68	328
						13.2						
4500	1.15	60	3.0	7.0	3.6	–	41	31	10	0.32	0.63	475
4600	1.16	58	3.0	7.0	3.6	–	28	14	7.5	0.53	0.68	351
4700	1.16	59	3.0	6.5	3.4	–	18.5	31	10.5	0.34	0.67	381

References

[1] P. Sun, Studies on Synthesis and Property of Mussel Mimetic Polyurethane, Shanghai Jiao Tong University, 2009.
[2] D. Wang, S. Gao, D. Dong, et al., A primary discussion on challenges for exploration and development of shale gas resources in China, Nat. Gas Ind. 33 (1) (2013) 8–17.
[3] J. Yuan, J. Deng, B. Wei, et al., Research on the wellbore stability of the horizontal wells of shale gas resources, Nat. Gas Ind. 32 (9) (2012) 66–70.
[4] G. Chen, M.E. Chenevert, M.M. Sharma, et al., A study of borehole stability in shales including poroelastic, chemical, and thermal effects, J. Pet. Sci. Eng. 38 (3) (2003) 167–176.
[5] L. Jiapeng, J. Zhen, Y. Bingye, et al., Structure and function of marine mussel adhesive proteins, Chin. J. Biochem. Mol. Biol. (1)(2007) 11–20.
[6] Q. Zhengsong, X. Jiafang, L. Kaihe, et al., A multivariate cooperation principle for well-bore stabilization, Acta Pet. Sin. 28 (2) (2007) 117–119.
[7] E. Van Oort, On the physical and chemical stability of shales, J. Pet. Sci. Eng. 38 (3) (2003) 213–235.
[8] K. Terzaghi, Theoretical Soil Mechanics, J. Wiley, New York, 1943.
[9] J.C. Jaeger, N.G. Cook, R. Zimmerman, Fundamentals of Rock Mechanics, John. Wiley & Sons, New York, 2009.
[10] H. Wensheng, Preparation, Structure and Properties of Silver-Loaded 4A Zeolite Antibacterial Agent and Silver-Zinc Nanometer SiO_2 Antibacterial Fiber, Taiyuan University of Technology, 2007.
[11] F. Yanru, Study on Optimum Extraction Process and Antitumor Mechanisms of Bifidobacterium Infantis Whole Peptidoglycan, Inner Mongolia Agricultural University, 2009.
[12] X. Fumin, The Pathogenic Role of Bacterial Flagellin in Ulcerative Colitis, Zhengzhou University, 2011.
[13] J. Jiang, L. Zhu, B. Zhu, et al., Surface characteristics of a self-polymerized dopamine coating deposited on hydrophobic polymer membranes, Langmuir 27 (23) (2011) 14180–14187.
[14] V. Ball, D. Del Frari, M. Michel, et al., Deposition mechanism and properties of thin polydopamine membranes for high added value applications in surface science at the nanoscale, BioNanoScience 2 (1) (2012) 16–34.
[15] L. Yang, S.L. Phua, J.K.H. Teo, et al., A bionic approach to enhancing interfacial interaction: polydopamine-coated clay as reinforcement for epoxy resins, ACS Appl. Mater. Interfaces 3 (8) (2011) 3026–3032.
[16] L. Cheng, J. Hu, F. Shen, et al., Bionic wellbore-strengthening drilling fluid technology for well Jia 3–21X, Drill. Fluid Completion Fluid 32 (6) (2015) 26–29. 104–105.
[17] G. Swan, Structure, chemistry, and biosynthesis of the melanins, Prog. Chem. Org. Nat. Prod. 31 (1974) 521.
[18] M.B. Clark Jr, J.A. Gardella Jr, T.M. Schultz, et al., Solid-state analysis of eumelanin biopolymers by electron spectroscopy for chemical analysis, Anal. Chem. 62 (9) (1990) 949–956.
[19] S. Subianto, Electrochemical Synthesis of Melanin-Like Polyindolequinone (Dissertation), The Queenslands University of Technology, 2006.
[20] G. Zajac, J. Gallas, J. Cheng, et al., The fundamental unit of synthetic melanin: a verification by tunneling microscopy of X-ray scattering results, Biochim. Biophys. Acta (BBA)-General Subj. 1199 (3) (1994) 271–278.
[21] P. Meredith, T. Sarna, The physical and chemical properties of eumelanin, Pigment. Cell Res. 19 (6) (2006) 572–594.

[22] E. Kang, K. Neoh, K. Tan, X-ray photoelectron spectroscopic studies of electroactive polymers, Polym. Charact. 106 (1993) 136–190.
[23] K.Y. Ju, Y. Lee, S. Lee, et al., Bioinspired polymerization of dopamine to generate melanin-like nanoparticles having an excellent free-radical-scavenging property, Biomacromolecules 12 (3) (2011) 625–632.
[24] A. Wilson, B. Kent, The glass-ionomer cement, a new translucent dental filling material, J. Appl. Chem. Biotechnol. 21 (11) (1971) 313.
[25] H. Zhuguo, Drilling Fluid Technique Study of Drilling Highly Deviated Well in Tremendous Thick Salt-Gypsum Formation, China University of Petroleum, 2007.
[26] M. Ritz, L. Vaculíková, E. Plevová, Application of infrared spectroscopy and chemometric methods to identification of selected minerals, Acta Geodyn. Geomater. 8 (1) (2011) 47–58.
[27] Z.Y. Xi, Y.Y. Xu, L.P. Zhu, et al., A facile method of surface modification for hydrophobic polymer membranes based on the adhesive behavior of poly (DOPA) and poly (dopamine), J. Membr. Sci. 327 (1–2) (2009) 244–253.
[28] L.Q. Xu, W.J. Yang, K.G. Neoh, et al., Dopamine-induced reduction and functionalization of graphene oxide nanosheets, Macromolecules 43 (2010) 8336–8339.
[29] Y. Hailong, Q. Ning, Research progress in waterborne polyurethane acrylate UV-curable oligomers, N. Chem. Mater. 39 (4) (2011) 51–54.
[30] L. Chuanhai, Research on the Technology of Water-Dispersed Amphoteric Polymer Polyacrylamide, Shandong University, 2007.
[31] L. Yao, H. Jinjun, L. Chunxia, D. Tianan, Development and performance evaluation of high-efficiency sulfonated phenolic resin, WestChina Explor. Eng. 22 (1) (2010) 43–44. 46.
[32] Y. Linqing, The Regularity and Mechanism Research of Inhibiting the Dispersion of Gangue, China University of Mining and Technology, 2014.
[33] M. Xueqin, Research and Application of Instability Mechanism and Preventive Technology of Side Deviated Well in Daqing Oilfield, China University of Petroleum, 2002.
[34] X. Tongtai, Z. Zhongju, X. Yunying, New advances of drilling fluid technology at home and abroad and the prospect of the 21st century (I), Drill. Fluid Completion Fluid (6)(2000) 33–40.
[35] C. Neuzil, How permeable are clays and shales? Water Resour. Res. 30 (2) (1994) 145–150.
[36] J. Randles, The real hydration energies of ions, Trans. Faraday Soc. 52 (1956) 1573–1581.
[37] H. Hefu, Application of electropositive polymer polyol drilling fluid in Linpan area, Drill. Fluid Completion Fluid 19 (4) (2002) 44–45.
[38] G. Feng, Application Research of Drilling and Completion Fluid Technology for Horizontal Well Protection Formation in Dagang Oilfield, China University of Petroleum, Beijing, 2008.
[39] Z. Xiaofeng, D. Xinlai, Technology and methods to improve well wall strength, China Well Rock. Salt 44 (5) (2013) 25–28.
[40] M.A. Ramirez, G. Sanchez, O.E. Preciado Sarmiento et al., Aluminum-based HPWBM successfully replaces oil-based mud to drill exploratory wells in an environmentally sensitive area, in: Proceedings of the SPE Latin American and Caribbean Petroleum Engineering Conference, Brazil, 2005.
[41] C. Bobko, F.-J. Ulm, The nano-mechanical morphology of shale, Mech. Mater. 40 (4) (2008) 318–337.
[42] Y. Liu, B. Yu, J. Hao, et al., Amination of surfaces via self-assembly of dopamine, J. Colloid Interface Sci. 362 (1) (2011) 127–134.

[43] P.F. Luckham, S. Rossi, The colloidal and rheological properties of bentonite suspensions, Adv. Colloid Interface Sci. 82 (1–3) (1999) 43–92.
[44] R.L. Anderson, I. Ratcliffe, H.C. Greenwell, P.A. Williams, S. Cliffe, P.V. Coveney, Clay expansion—a challenge in the oilfield, Earth Sci. Rev. 98 (3–4) (2010) 201–216.
[45] E. Kienzl, L. Puchinger, K. Jellinger, et al., The role of transition metals in the pathogenesis of Parkinson's disease, J. Neurol. Sci. 134 (Suppl.) (1995) 69–78.
[46] P. Cunningham, Dif Holdfasts. Biological Adhesives, 2006:125–143. Ferential Thermal Analysis, Academic Press, London, 1970.
[47] H. Lee, S.M. Dellatore, W.M. Miller, et al., Mussel-inspired surface chemistry for multifunctional coatings, Science 318 (5849) (2007) 426–430.
[48] S. Fangmin, Z. Jupeng, D. Xun, et al., The preparation of PVA/quaternized chitosan composite hydrogel and its application as artificial dressing for burn wound healing, J. Funct. Mater. 46 (9) (2015) 9133–9138. 9143.
[49] K. Xiangling, M. Ohadi, Applications of micro and nano technologies in the oil and gas industry - overview of the recent progress, in: Proceedings of the Abu Dhabi International Petroleum Exhibition and Conference, Abu Dhabi, 2010.
[50] B. Wei, Z. Wang, B. Guo, Theoretical research and progress of wellbore instability in mud shale formations, Drill. Prod. Technol. 30 (3) (2007) 16–20.
[51] H. Rongzun, C. Mian, Study on shale stability of wellbore by mechanics coupling with chemistry method, Drill. Fluid Completion Fluid 12 (3) (1995) 15–21.

CHAPTER 2

New technology of superlubricated bionic drilling fluids

In recent years, because of diminishing shallow underground oil reserves, increasing efforts have been made to drill complex wells such as extended reach wells, horizontal wells, deep wells, and ultradeep wells. In the process, the contradiction between the lubricity of drilling fluids and downhole safety has become increasingly obvious [1]. The process of drilling deep wells and extended reach wells, owing to the high friction between drilling tool, borehole, and casing, is often prone to complicated accidents such as frictional resistance to the round trip and pipe sticking during drilling. Furthermore, the high friction not only generates all kinds of friction and wear that cause equipment damage, but also brings high torque and high annulus circulation pressure consumption, thereby greatly wasting the energy of the drilling rig and reducing the efficiency of rock breaking. At present, most drilling operations in China incorporate water-based drilling fluids. However, compared with oil-based drilling fluids, the commonly used water-based drilling fluids have higher frictional resistance and cannot meet the lubrication requirements for drilling complex wells, such as deep and horizontal wells. Adjusting the lubricity of a water-based drilling fluid by adding lubricants is one of the main technical means of preventing and solving drilling safety problems and improving drilling efficiency. The added lubricants play a key role in reducing the friction between the drilling tool and the borehole as well as reducing the drilling torque and frictional resistance during the round trip [2].

However, in addition to the problems of easy foaming, poor compatibility with drilling fluids, and environmental pollution, the lubricants that are commonly used in water-based drilling fluids almost all work by changing the characteristics and flow patterns of drilling fluids, which cannot form a smooth film between the drilling tool and the borehole, leading to poor lubrication effects. Therefore there is an urgent need to develop a highly efficient and environmentally friendly drilling fluid

Fundamentals and Applications of Bionic Drilling Fluids
DOI: https://doi.org/10.1016/B978-0-323-90293-9.00002-2

lubricant that can meet the requirements of safe, fast, and efficient operation during drilling and does not have a negative impact on other properties of the drilling fluid, such as rheological properties and filtration loss.

In this chapter, bionics is introduced into the field of drilling fluids to develop a high-performance bionic bonded water-based drilling fluid lubricant. The active component in the lubricant associates with free ions in the wellbore to produce a smooth film on the surface of the drilling tool and the borehole, effectively controlling the inherent vortex in the flow interface, reducing the frictional resistance in the drilling of complex wells such as the deviated section of extended reach wells, improving the mechanical penetration rate, extending the life of drilling tools, and reducing the operational costs. In addition, the bonded lubricant can improve the rheological properties and reduce the filtration loss of the drilling fluid system and is environmentally friendly, which effectively improves the overall benefits of drilling. Based on the developed bonded lubricant and the principle of flow and bonded lubrication, a new technology of anti-pipe-sticking drilling fluid with the synergistic effect of flow and bonded lubrication has been established and has good application prospects.

2.1 Bionic bonded water-based drilling fluid lubricants based on the principle of the viscosity and resistance reduction of earthworms

2.1.1 Analysis of the principle of earthworms' viscosity and resistance reduction and its bionic application in drilling fluids

Earthworms are typical soil animals. In the environment where soil animals live, the soil water content is usually between the plastic limit and the flow limit, making it easy to adhere to foreign objects. But soil animals can move freely in such soil, and soil does not adhere to their bodies. This is mainly the result of the good viscosity and resistance reduction performance of their body surface. Studies have shown that the viscosity and resistance reduction properties of earthworms in soil are derived mainly from the strong hydrophobicity of the body surface and the composition of their surface which is beneficial to viscosity and resistance reduction. That is, in addition to carbon, hydrogen, oxygen, and nitrogen, earthworms' surface contains trace elements such as silicon, phosphorus, calcium, chlorine, potassium, sulfur, iron, aluminum, and manganese, which exist mainly in the form of inorganic salts. These trace elements are beneficial for reducing viscosity and resistance.

During the drilling process, the movement of drill bit and drilling tool in the formation is very similar to the movement of soil animals in the soil. To improve the penetration rate and the quality of the wellbore, great efforts should be made to reduce the frictional resistance between the drilling tool, drilling fluid, and borehole; reduce the adhesion of clay to the drill bit and drilling tool; reduce the energy consumption resulting from friction; and ensure that the mud pump has the power to break the rock. Especially in the drilling of coalbed methane, shale oil and gas, tight oil and gas, and complex structure wells, the requirement for the lubricity of drilling fluid must be higher.

For this reason high-performance lubricants are added to drilling fluids. Currently, lubricants are classified as solid particles or liquids. The mechanism of solid particle lubricants is their rolling on the borehole. By contrast, liquid lubricants typically change the characteristics and flow patterns of drilling fluids, where there is no bonding effect between the lubricant and the borehole or metal. Thus the friction between the drilling too, drilling fluid, and borehole is still large; the adhesion of the drill bit to cuttings is strong; and the phenomenon of bit baling is serious, which will bring negative effects to the rheological properties of drilling fluids. In response to these problems, the principle of earthworms' viscosity and resistance reduction was introduced in order to develop an efficient lubricant that can produce a bonding effect with the borehole and the surface of the drilling tool, effectively controlling the inherent vortex in the flow interface, reducing the frictional resistance in the drilling of complex wells such as the deviated section of extended reach wells, improving the mechanical penetration rate, extending the life of drilling tools, reducing the operational costs, and effectively improving the overall benefits of drilling.

2.1.2 Development and performance evaluation of a bionic bonded lubricant

2.1.2.1 Development of the bonded lubricant

At present, almost all liquid lubricants work by changing the characteristics and flow patterns of drilling fluids. However, the lubrication effect is not ideal, and it will have an adverse effect on the rheological properties of drilling fluids. This type of lubricant will not bond with the drilling tool and borehole, will not form a smooth and dense film on the surface of the drilling tool and borehole, and will not prevent the occurrence of an inherent vortex in the flow interface of the drilling fluid. Therefore the frictional and flow resistance of the drilling fluid are still very large in the circulation process.

To solve these problems, a bonded drilling fluid lubricant was developed (Fig. 2.1) based on the principle of earthworms' viscosity and resistance reduction in soil and characterized by infrared technique. As shown in Fig. 2.2, 2924.9 cm^{-1} is attributed to the stretching vibration of methyl-CH_3−, and 2853.6 cm^{-1} reflects the stretching vibration of methylene-CH_2−. 1741.9 cm^{-1} is the characteristic peak of $-C=O$ in ester bond-COO−, and 1463.9 cm^{-1} is the deformation vibration absorption peak of methyl-CH_3−. 1168.4 cm^{-1} is the vibration absorption peak of the single bond C−O in the ester bond. There are no absorption peaks of the carboxylic acid and hydroxyl group, indicating relatively complete esterifications.

The active component in the lubricant associates with free ions in the wellbore to produce a smooth film on the surface of the drilling tool and the borehole, which effectively controls the inherent vortex in the flow interface, reduces the frictional resistance in the drilling of complex wells such as the deviated section of extended reach wells, improves the mechanical penetration rate, extends the life of drilling tools, reduces the operational costs, and effectively improves the overall benefits of drilling. Based on the developed bonded lubricant and the principle of flow and

Figure 2.1 The bonded drilling fluid lubricant.

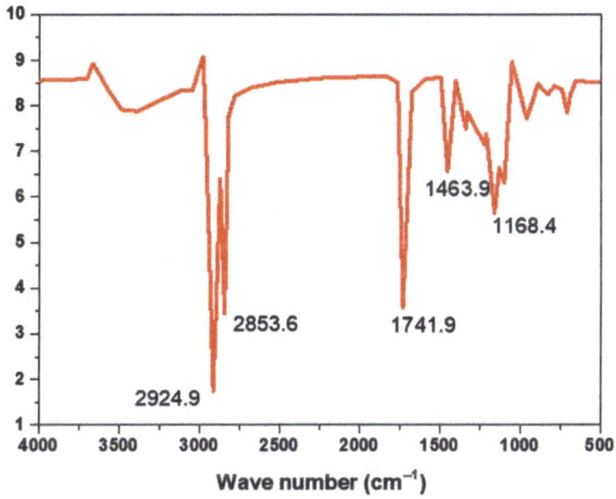

Figure 2.2 Infrared spectrum of the bonded drilling fluid lubricant.

bonded lubrication, a new technology of anti-pipe-sticking drilling fluid with the synergistic effect of flow and bonded lubrication has been established.

2.1.2.2 Performance evaluation of the bionic bonded lubricant

2.1.2.2.1 Evaluation of lubricity in pure water

First, 1% lubricant was added in pure water and stirred for 20 min. Then an extreme pressure (EP) lubricator was used to measure its lubrication coefficient. After measuring, each sample was put in the aging tank and hot rolled at 150°C for 16 h. After the hot rolling was finished, the sample was taken out and stirred for 20 min. The lubrication coefficient of each sample was measured again. Its lubrication effect was evaluated and compared with international advanced products. The results are shown in Table 2.1. Before and after the hot rolling, the lubrication coefficient of these lubricant products in pure water can all be significantly reduced.

2.1.2.2.2 Evaluation of lubricity in 4% base mud

First, 40 g of bentonite and 3 g of anhydrous sodium carbonate were added to 1 L of water, stirred at high speed for 5 h, and set aside for 24 h at room temperature to obtain fresh water-based mud with a soil content of 4%. A certain amount of 4% base mud was taken, and 1% of each lubricant product was added to it, followed by stirring at high speed for 20 min and then measuring of the lubrication coefficient. After hot rolling at 150°C for 16 h, the lubrication coefficient was measured again, with

Table 2.1 Lubricity of lubricant in pure water.

Sample	EP lubrication coefficient	Reduction rate of lubrication coefficient
Pure water	0.36	—
Pure water + 1% international advanced product (GRANDOIL Company)	0.06	83.33%
Pure water + 1% international advanced product (DFL)	0.03	91.67%
Pure water + 1% bonded lubricant	0.04	88.89%
Sample (hot rolling at 150°C for 16 h)	**EP lubrication coefficient**	**Reduction rate of lubrication coefficient**
Pure water	0.36	—
Pure water + 1% international advanced product (GRANDOIL Company)	0.07	80.56%
Pure water + 1% international advanced product (DFL)	0.05	86.11%
Pure water + 1% bonded lubricant	0.05	86.11%

the results shown in Table 2.2. The lubrication effect of domestic lubricant products in 4% base mud was unremarkable or very poor. Among them, the relatively good lubricant CX-300H had a serious foaming problem. By contrast, the bonded lubricant and the international advanced DFL lubricant showed excellent lubricity, with a similar performance of reducing the lubrication coefficient in 4% base mud, and the reduction rates were all above 80%. Notably, the bonded lubricant was slightly better than the international advanced DFL lubricant.

Except for Zhanjiang PF-lube lubricant, whose lubricity remained basically unchanged, the lubricity of the other lubricants improved After hot rolling at 150°C, and the lubrication coefficient in 4% base mud was reduced by more than 70%. The GRANDOIL Company lubricant and the DFL lubricant showed a reduction rate in the lubrication coefficient close to 80%, while the bonded lubricant had the best lubrication effect in the base mud, the reduction rate of its lubrication coefficient reaching 90%. In summary, in terms of the reduction in the lubrication coefficient in base mud, the bonded lubricant is superior to the existing excellent lubricants in the world and solves the foaming problem that commonly exists in China's lubricants.

Table 2.2 Lubricity of lubricants in base mud.

Sample	EP lubrication coefficient	Reduction rate of lubrication coefficient	Foaming phenomenon
4% Base mud	0.54	—	No foaming
4% Base mud + 1% Chinese advanced product (PF-lube)	0.32	40.7%	Slight foaming
4% Base mud + 1% Chinese advanced product (CX-300H)	0.16	70.4%	Serious foaming
4% Base mud + 1% Chinese advanced product (PF-lube)	0.31	42.6%	Slight foaming
4% Base mud + 1% international advanced product (GRANDOILCompany)	0.22	59.3%	Slight foaming
4% Base mud + 1% international advanced product (DFL)	0.10	81.5%	Slight foaming
4% Base mud + 1% bonded lubricant	0.09	83.3%	Slight foaming
Sample (hot rolling at 150°C for 16 h)	**EP lubrication coefficient**	**Reduction rate of lubrication coefficient**	**Foaming phenomenon**
4% Base mud	0.52	—	No foaming
4% Base mud + 1% Chinese advanced product (PF-lube)	0.32	38.5%	Obvious foaming
4% Base mud + 1% Chinese advanced product (CX-300H)	0.15	71.2%	Serious foaming
4% Base mud + 1% Chinese advanced product (PF-lube)	0.13	75.0%	Obvious foaming
4% Base mud + 1% international advanced product (GRANDOILCompany)	0.11	78.8%	Slight foaming
4% Base mud + 1% international advanced product (DFL)	0.08	84.6%	Slight foaming
4% Base mud + 1% bonded lubricant	0.05	90.4%	Almost no foaming

2.1.2.2.3 Lubrication effect evaluation for different dosages of lubricant

By analyzing the effect of lubricant dosage on the lubrication effect, a relatively reasonable dosage can be determined. The results are shown in Fig. 2.3. After 0.2% bonded lubricant was added to 4% base mud at room

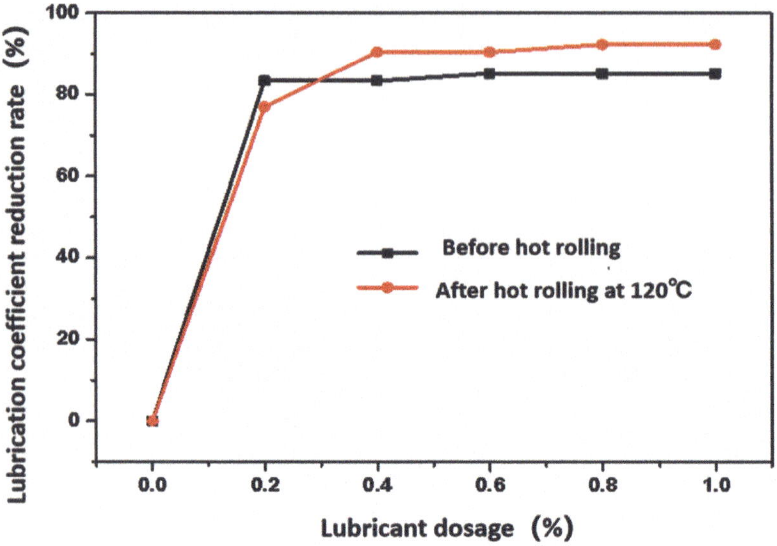

Figure 2.3 Effect of lubricant dosage.

temperature, a good lubrication effect was achieved that reduced the lubrication coefficient by more than 80%. Further increasing the lubricant dosage did not significantly improve the lubrication effect. After hot rolling at 120°C, the lubrication effect with 0.2% dosage was reduced, owing to the decomposition of a small amount of lubricant, but when the lubricant dosage was 0.4%, the lubrication coefficient was reduced by about 90%. Further increasing the dosage improved the lubrication effect slightly but not significantly. Considering that there are various agents and a large amount of barite in the drilling fluid system, the optimal dosage of the lubricant is 1.5%, and the dosage can be increases appropriately for systems with a higher density.

2.1.2.2.4 Evaluation of the reduction rate of flow resistance of base mud

When fluid flows in a pipeline, resistance is generated owing to the effects of viscosity and vortex. Therefore a loop flow resistance test system (Fig. 2.4) in the laboratory was used to evaluate the effect of the bonded lubricant on the flow resistance of base mud.

The experimental instrument consisted of a loop flow resistance test device containing two 3-m-long test tubes, and the pressure measurement point was set in the middle of the 2.5 m section by trepanning. The

Figure 2.4 Flow resistance test system.

experimental power source was a screw pump, and a mass flow meter was installed downstream of the outlet end of the test tube. Through the feedback of a digital flow signal, the fluid displacement could be adjusted according to the actual demand.

Because of the significant influence of the pipe diameter on the experimental results, fresh water was used to calibrate the pipe diameter [3–7]. Based on the existing benchmark data of pure water, the flow of drilling fluid was simulated, and the pressure drop of the base mud with different lubricant dosages at the same displacement was measured. The average value was calculated by measuring for multiple times and was then compared with the value for the blank group to calculate the reduction rate, as shown in Fig. 2.5. Initially, as the dosage of bonded lubricant increased, the reduction rate of flow resistance also increased. When the amount of bonded lubricant was 0.5%, the flow resistance of the drilling fluid could be reduced by 12.5%. Further increasing the amount of lubricant did not lead to significant change in the reduction rate of flow resistance.

2.1.2.2.5 Evaluation of the cake adhesion coefficient in base mud

During the drilling process, various types of pipe sticking often occur, such as reduced diameter sticking and sand bridge sticking. The most common and most serious type of pipe sticking is the sticky pipe sticking of cake (also called differential pressure sticking or adhesion sticking). The existence of formation pore pressure and drilling fluid column pressure is the external reason for the formation of sticky pipe sticking; the internal reason is the existence of cake on the borehole [8–10]. The greater the friction coefficient of the cake, the greater is the adhesion of the drill string to the cake and the torque value. To prevent the occurrence of

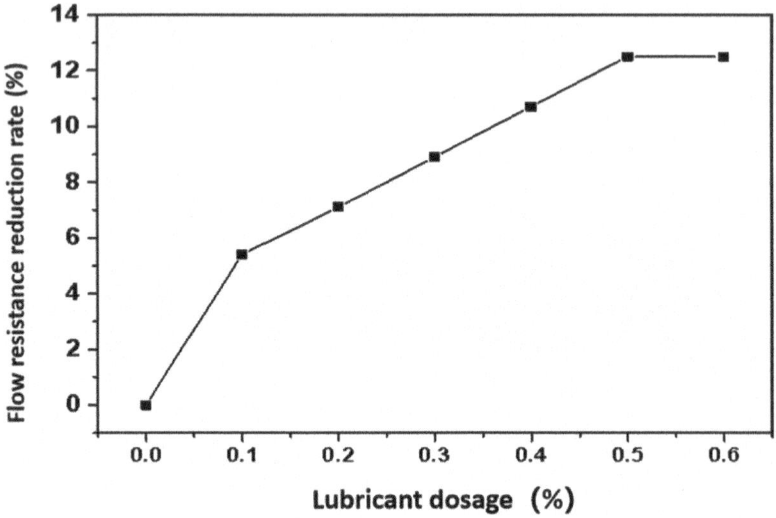

Figure 2.5 Effect of bonded lubricant on the flow resistance of drilling fluid.

such pipe-sticking accidents, the cake adhesion coefficient of mud must be measured frequently during drilling [11–13].

The bonded lubricant and several other lubricants were added to the 4% base mud, followed by being pressurized and filtered for 30 min on the adhesion coefficient tester, after which the cake adhesion coefficient was measured. After hot rolling at 120°C for 16 h, the same procedure was repeated, and the relevant data were measured again. The experimental results are shown in Table 2.3. Before hot rolling, the lubricants of CX-300H, DFL, and GRANDOIL Company and the bonded lubricant all reduced the cake adhesion coefficient by more than 50%; the DFL lubricant had the strongest ability to reduce the adhesion coefficient. After the hot rolling, the adhesion coefficient of most lubricants was further reduced, and the best effect was still achieved by the international advanced DFL lubricant. Generally speaking, the bonded lubricant is close to the internationally advanced lubricants in terms of reducing the adhesion coefficient.

2.1.2.2.6 Experimental evaluation of wear resistance

The four-ball friction tester is usually used to determine the long-term wear resistance of lubricants under specified conditions, and the test results are analyzed based on the friction coefficients and scratches. Three of the four balls are clamped into a nest and placed in the lower cup, and the

Table 2.3 Cake adhesion coefficients.

Sample	Before hot rolling		After hot rolling	
	Adhesion coefficient	Reduction rate	Adhesion coefficient	Reduction rate
4% Base mud	0.1225	—	0.1098	—
4% Base mud + 1% Chinese advanced product (PF-lube)	0.0845	31.0%	0.0676	38.5%
4% Base mud + 1% Chinese advanced product (CX-300H)	0.0592	51.7%	0.0592	46.2%
4% Base mud + 1% Chinese advanced product (PF-lube)	0.0676	44.8%	0.0465	57.7%
4% Base mud + 1% international advanced product (GRANDOIL Company)	0.0592	51.7%	0.0448	59.2%
4% Base mud + 1% international advanced product (DFL)	0.0465	62.1%	0.0380	65.4%
4% Base mud + 1% bonded lubricant	0.0549	55.2%	0.0465	57.7%

other ball is placed above. Under axial load, the upper ball is in point contact with the other three balls. The upper ball is driven to rotate to produce relative sliding with the other three balls below.

The 0.5% bonded lubricant and the other advanced lubricants were added in pure water. Pure water alone was a blank group. A four-ball friction tester was used to grind for 30 min under the load of 150 N and a rotation speed of 100 rpm. Then a field emission environment scanning electron microscope (FE-ESEM, Quanta 200F) was used for observing and analyzing the scratches on the steel ball after long grinding. The experimental results are presented as follows.

As can be seen from Fig. 2.6, in the wear spot of pure water there are obvious scratches and deep furrows, and the area of the wear spot is the largest. After the lubricant was added, the diameter of the wear spot was reduced. The scratches with the bonded lubricant are the least obvious, and the friction surface is the smoothest; but after other lubricants are added, there are still more obvious scratches.

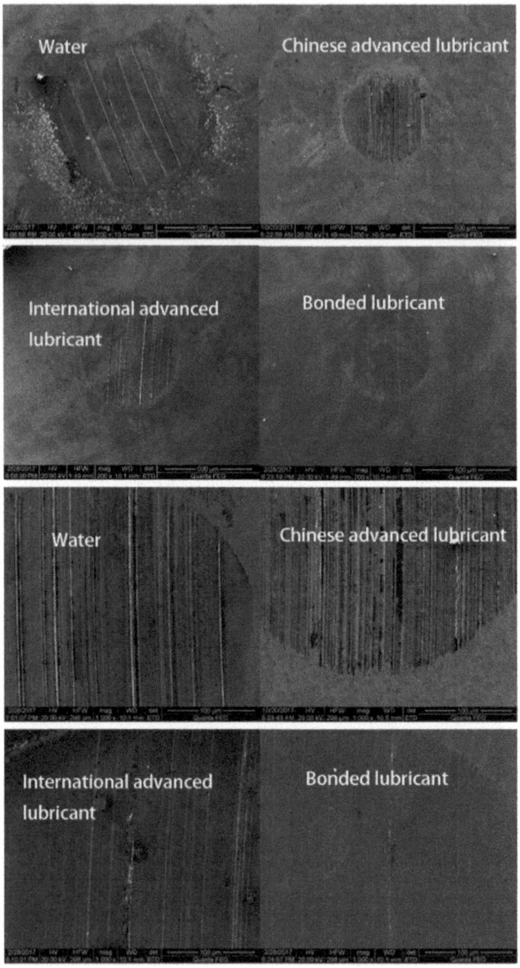

Figure 2.6 Scratches obtained by the four-ball friction tester.

2.1.2.2.7 Evaluation of salt resistance

The 1% bonded lubricant was added to the 4% base mud to set a blank group, and a certain amount of NaCl, $CaCl_2$, and KCl was added as control groups. The extreme pressure lubrication coefficient of each group before and after hot rolling was measured to evaluate the salt resistance capacity of the bonded lubricant, with the results shown in Table 2.4. Compared with the blank group, after the addition of 20% NaCl, 30% $CaCl_2$, or 10% KCl at room temperature, the performance of the bonded lubricant was not significantly affected. When 30% NaCl or 10% or more

Table 2.4 Salt resistance test.

Sample	Lubrication coefficient	Reduction rate of lubrication coefficient	Notes
4% Base mud	0.52	—	
4% Base mud + 1% lubricant	0.08	84.6%	
4% Base mud + 10% NaCl + 1% lubricant	0.08	84.6%	
4% Base mud + 20% NaCl + 1% lubricant	0.08	84.6%	
4% Base mud + 30% NaCl + 1% lubricant	0.09	82.7%	Slight foaming
4% Base mud + 10% $CaCl_2$ + 1% lubricant	0.08	84.6%	
4% Base mud + 20% $CaCl_2$ + 1% lubricant	0.09	82.7%	
4% Base mud + 30% $CaCl_2$ + 1% lubricant	0.08	84.6%	
4% Base mud + 5% KCl + 1% lubricant	0.08	84.6%	
4% Base mud + 10% KCl + 1% lubricant	0.08	84.6%	Slight foaming
4% Base mud + 20% KCl + 1% lubricant	0.10	80.8%	Slight foaming
Sample (after hot rolling at 120°C for 16 h)	**Lubrication coefficient**	**Reduction rate of lubrication coefficient**	
4% Base mud	0.51	—	
4% Base mud + 1% lubricant	0.05	90.2%	
4% Base mud + 10% NaCl + 1% lubricant	0.05	90.2%	
4% Base mud + 20% NaCl + 1% lubricant	0.06	88.2%	
4% Base mud + 30% NaCl + 1% lubricant	0.07	86.3%	
4% Base mud + 10% $CaCl_2$ + 1% lubricant	0.05	90.2%	
4% Base mud + 20% $CaCl_2$ + 1% lubricant	0.06	88.2%	
4% Base mud + 30% $CaCl_2$ + 1% lubricant	0.10	80.4%	
4% Base mud + 5% KCl + 1% lubricant	0.07	86.3%	
4% Base mud + 10% KCl + 1% lubricant	0.08	84.3%	Slight foaming
4% Base mud + 20% KCl + 1% lubricant	0.07	86.3%	Slight foaming

KCl was added, there was a certain amount of foaming, but it was within the acceptable range, and the lubricity was not greatly affected. After hot rolling, there was a slight foaming phenomenon after KCl was added. After 30% $CaCl_2$ was added, the lubricity was slightly reduced, but the reduction rate of the lubrication coefficient was still above 80%, and those of most others' lubrication coefficients were still higher than 85%. In general, the bonded lubricant has good salt resistance, and it can maintain its good lubricity after a large amount of salt is added.

Table 2.5 Temperature resistance test.

Hot rolling temperature	Sample	Lubrication coefficient	Reduction rate	Notes
Room temperature	4% Base mud	0.54	—	
	4% Base mud + 1% bonded lubricant	0.09	83.3%	
Hot rolling at 120°C for 16 h	4% Base mud	0.52	—	
	4% Base mud + 1% bonded lubricant	0.04	92.3%	
Hot rolling at 150°C for 16 h	4% Base mud	0.50	—	
	4% Base mud + 1% bonded lubricant	0.04	92.0%	
Hot rolling at 180°C for 16 h	4% Base mud	0.50	—	
	4% Base mud + 1% bonded lubricant	0.05	90.0%	
Hot rolling at 200°C for 16 h	4% Base mud	0.48	—	
	4% Base mud + 1% bonded lubricant	0.07	85.4%	
Hot rolling at 220°C for 16 h	4% Base mud	0.45	—	
	4% Base mud + 1% bonded lubricant	0.07	84.4%	
Hot rolling at 240°C for 16 h	4% Base mud	0.42	—	
	4% Base mud + 1% bonded lubricant	0.09	78.6%	Slight foaming

2.1.2.2.8 Evaluation of temperature resistance

The 1% bonded lubricant was added to 4% base mud and was hot rolled at different temperatures for 16 h. The lubrication coefficient of the hot-rolled base mud was measured and compared to evaluate its temperature resistance. The test results are shown in Table 2.5. After hot rolling at 120°C the reduction rate in the lubrication coefficient of the bonded lubricant increased; as the hot rolling temperature increased, the reduction rate began to decrease slightly. After hot rolling at 180°C the reduction rate of the lubrication coefficient was still around 90%; at a hot rolling temperature of 220°C the reduction rate of the lubrication coefficient was still around 85%. Therefore the bonded lubricant has a strong temperature resistance of at least 180°C.

2.1.2.2.9 Evaluation of wettability

Under boundary lubrication conditions, the lubricating film is extremely thin, and the adsorption performance of the lubricant has a great influence

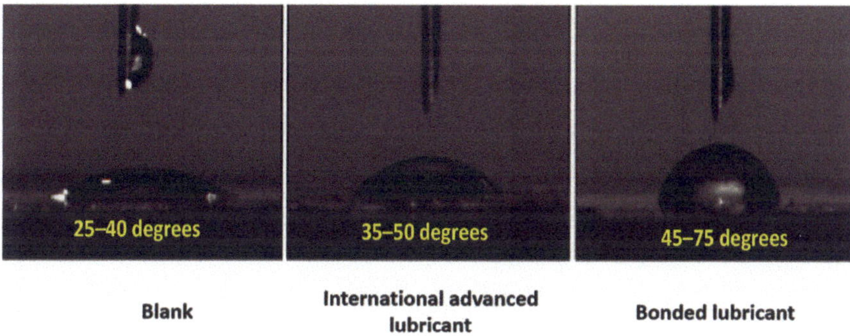

Figure 2.7 Water contact angle on the glass substrate.

on the lubricity. The adhesion energy on which lubricant molecules rely to form an adsorption film on the friction surface is closely related to its surface wettability. The adsorption ability of lubricant molecules on the solid surface can be judged indirectly through the change of the wettability of the solid surface, which can be evaluated by measuring the contact angle of the liquid on the solid surface.

The glass and steel substrates were immersed in the aqueous solution of 1% bonded and other advanced lubricants for 24 h. They were then taken out and dried in a drying box, and their water contact angle was then measured by using a contact angle measuring instrument. The results were compared with those of the blank group. The change in the wettability of the surface of the glass and steel substrates was determined in order to determine the adsorption capacity of the lubricant on the solid surface.

Fig. 2.7 shows the water contact angle of the glass substrates treated with different lubricants. All the lubricants had a certain amount of adsorption on the glass surface. The international advanced lubricant could slightly increase the hydrophobicity of the surface of the glass substrate, while the effect of the bonded lubricant was more obvious, indicating that the bonded lubricant was easier to adsorb on the glass surface than the others to form an adsorption film.

The water contact angle of the steel substrates treated with different lubricants is shown in Fig. 2.8. Both of the lubricants could enhance the hydrophobicity of the steel substrate surface, indicating that both lubricants could form a strong adsorption film on the surface of the steel substrate. However, the performance of the bonded lubricant is better, and the water contact angle is significantly greater than that of the international advanced lubricant.

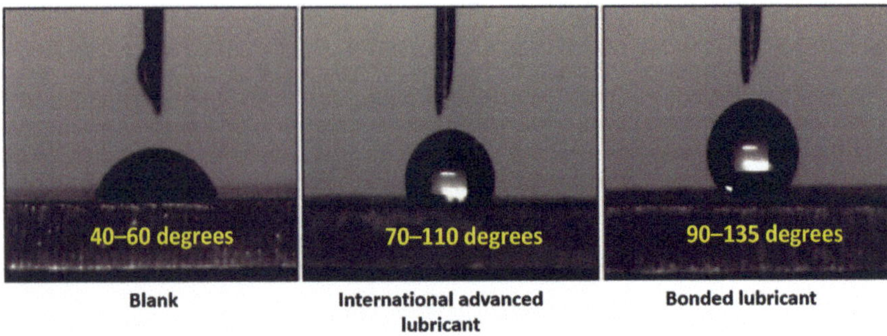

Figure 2.8 Water contact angle on the steel substrate.

In summary, the bonded lubricant is more easily adsorbed on boreholes, casings, and downhole tools to form a lubricating film than is the case for the international advanced lubricant, which shows better lubricity.

2.1.2.2.10 Evaluation of environmental friendliness

Biological toxicity is evaluated according to China's standard SY/T 6788-2010 of Evaluation Procedures of Environmental Protection for Water-Soluble Oilfield Chemicals. A 2% bonded lubricant/distilled water solution was prepared as the test solution. One sample of test solution was taken according to volume ratio and added to nine samples of 30 g/L sodium chloride solution, followed by stirring at 11,000 rpm for 30 min and then standing for 60 min. The concentration of this sample solution was set to 106 mg/L, and it was then diluted to different concentrations of sample solutions. Then tests were conducted by using a bioluminescence photometer and luminescent bacteria. The relative luminosity was measured and calculated, curves were drawn, and the corresponding concentrations were obtained. The final measurement results are as follows: Luminescence bacteria EC50 > 106 mg/L. According to China's standard SY/T 6788-2010 of Evaluation Procedures of Environmental Protection for Water-Soluble Oilfield Chemicals, the biotoxicity of EC50 > 20,000 mg/L is nontoxic, so the bonded lubricant has no biological toxicity and meets the requirements for environmental protection.

2.1.2.2.11 Evaluation in drilling fluid system

1. Evaluation in polysulfonate system

   The bonded lubricant, the DFL lubricant, and the GRANDOIL lubricant were added to the system and compared with the

Table 2.6 Lubrication performance in a polysulfon system.

Room temperature	AV (mPa · s)	PV (mPa · s)	YP (Pa)	Gel (Pa/Pa)	FL (API) (mL)	Cake sluggish coefficient
System 1	84	76	8	1.5/5	—	—
System 2	75	66	9	2/6	—	—
System 3	73	6	8.2	2/5.5	—	—
System 4	72.5	66	6.6	2/6	—	—
Hot rolling at 150°C for 16 h	**AV (mPa · s)**	**PV (mPa · s)**	**YP (Pa)**	**Gel (Pa/Pa)**	**FL (API) (mL)**	**Cake sluggish coefficient**
System 1	76	64	12.3	10/20	11.2	0.212
System 2	74.5	64	10.7	6/14	2.8	0.105
System 3	69	58	11.2	4/13	3.5	0.123
System 4	72	62	10.2	2.5/10	2.4	0.096

blank group to measure and calculate the apparent viscosity (AV), plastic viscosity (PV), yield point/plastic viscosity (YP/PV) ratio, and filtration loss before and after the hot rolling to perform a comprehensive evaluation of lubricant performance, as shown in Table 2.6.

System 1: 2% bentonite + 1% polymer + 4% SMP − II + 2% FT − 1 + 1% RSTF + 0.2% NaOH + barite, $\rho = 2.2\ \text{g/cm}^3$.

System 2: 2% bentonite + 1% polymer + 4% SMP − II + 2% FT − 1 + 1% RSTF + 0.2% NaOH + 2% GRANDOIL lubricant + barite, $\rho = 2.2\ \text{g/cm}^3$.

System 3: 2% bentonite + 1% polymer + 4% SMP − II + 2% FT − 1 + 1% RSTF + 0.2% NaOH + 2% DFL lubricant + barite, $\rho = 2.2\ \text{g/cm}^3$.

System 4: 2% bentonite + 1% polymer + 4% SMP − II + 2% FT − 1 + 1% RSTF + 0.2% NaOH + 2% bonded lubricant + barite, $\rho = 2.2\ \text{g/cm}^3$.

The experimental results show that before hot rolling, the viscosity of each system decreased after the lubricants were added; after hot rolling, the GRANDOIL lubricant (system 2) did not significantly reduce the viscosity but reduced the YP, while the DFL (system 3) and bonded lubricant (system 4) reduced the viscosity and YP of the system, thus improving the rheological properties of the system.

In addition, after the lubricants were added, the API filtration loss after hot rolling of each system was significantly reduced, and the sluggish coefficient of the cake was significantly reduced. The most obvious effect was achieved by the bonded lubricant, with only 2.4 mL filtration loss and a 54.7% decrease in the cake sluggish coefficient. This is because the bonded lubricant could form a low-permeability hydrophobic oil film on the surface of the cake.

2. Evaluation in soilless KCl system

System 1: Water + 0.1% KPAM + 3% starch + 3% potassium humate + 3% RSTF + 2% white oil + 0.5% OP − 4 + 3% white asphalt + 1% inhibitor + 8% KCl, $\rho = 2.4\ g/cm^3$.

System 2: Water + 0.1% KPAM + 3% starch + 3% potassium humate + 3% RSTF + 2% white oil + 0.5% OP − 4 + 3% white asphalt + 1% inhibitor + 8% KCl + 2% GRANDOIL lubricant, $\rho = 2.4\ g/cm^3$.

System 3: Water + 0.1% KPAM + 3% starch + 3% potassium humate + 3% RSTF + 2% white oil + 0.5% OP − 4 + 3% white asphalt + 1% inhibitor + 8% KCl + 2% DFL lubricant, $\rho = 2.4\ g/cm^3$.

System 4: Water + 0.1% KPAM + 3% starch + 3% potassium humate + 3% RSTF + 2% white oil + 0.5% OP − 4 + 3% white asphalt + 1% inhibitor + 8% KCl + 2% bonded lubricant, $\rho = 2.4\ g/cm^3$.

The experimental results are shown in Table 2.7. At normal temperature the addition of the GRANDOIL lubricant (system 2) had no obvious viscosity-reducing effect but instead increased the viscosity. The addition of the DFL lubricant (system 3) and the bonded lubricant (system 4) had the obvious effects of reducing viscosity and gel strength, indicating that these two lubricants could be adsorbed on the surface of barite at normal temperature and reduce the friction between particles. After hot rolling at 130°C for 16 h, each lubricant showed a certain degree of viscosity-reducing effect. The DFL lubricant (system 3) had the most obvious effect in reducing viscosity and gel strength. The GRANDOIL lubricant (system 2) and the bonded lubricant (system 4) had basically the same effect.

Table 2.7 Lubrication performance in a soilless KCl system.

System	AV (mPa · s)	PV (mPa · s)	YP (Pa)	Gel (Pa/Pa)	FL (API) (mL)
System 1	115.5	76	39.5	28/54	
System 2	134	106	28	16/55	
System 3	83	65	18	9/36	
System 4	81.5	64	17.5	9/34	
Hot rolling at 130°C for 16 h	**AV (mPa · s)**	**PV (mPa · s)**	**YP (Pa)**	**Gel (Pa/Pa)**	**FL (API) (mL)**
System 1	105	95	10	2.5/20.5	2.5
System 2	89	76	13	2/14	3.9
System 3	69.5	65	4.5	1/10	2.7
System 4	89	78	11	1.5/9.5	2.4

Table 2.8 Environmentally friendly system with a density of 1.6 g/cm^3.

Room temperature	AV (mPa · s)	PV (mPa · s)	YP (Pa)	Gel (Pa/Pa)	FL (API) (mL)	HTHP (API) (mL)	Cake sluggish coefficient
System 1	78.5	54	25.04	2.5/13	5		
System 2	80.5	61	19.93	2.5/9.5	3.4		
Hot rolling at 120°C for 16 h	**AV (mPa · s)**	**PV (mPa · s)**	**YP (Pa)**	**Gel (Pa/Pa)**	**FL (API) (mL)**	**HTHP (API) (mL)**	**Cake sluggish coefficient**
System 1	113	96	17.37	2.5/7	4.6	18	0.112
System 2	88	70	18.40	2/4	4.8	12	0.052

Overall, in the soilless KCl system, the lubricating and viscosity-reducing effects of the bonded lubricant are similar to those of the GRANDOIL lubricant but are still weaker than those of the DFL lubricant.

3. Evaluation in environmentally friendly drilling fluid systems
   a. The experimental results of environmentally-friendly drilling fluid systems with density of 1.6 g/cm^3 are shown in Table 2.8

System 1: 400 mL water + 2% bentonite + 0.3% NaOH
+ 0.5% filtration control agent + 1% silicone viscosity reducer
+ 1% YP increaser + 0.5% amine-inhibitor
+ 4% ultrafine calcium + 350 g barite

Table 2.9 Environmentally friendly system with a density of 1.8 g/cm^3.

Room temperature	AV (mPa · s)	PV (mPa · s)	YP (Pa)	Gel (Pa/Pa)	FL (API) (mL)	HTHP (API) (mL)	Cake sluggish coefficient
System 1	83	60	23.51	2.5/8.5	3.4		
System 2	82.5	61	21.97	2/7	3.2		
Hot rolling at 120°C for 16 h	**AV (mPa · s)**	**PV (mPa · s)**	**YP (Pa)**	**Gel (Pa/Pa)**	**FL (API) (mL)**	**HTHP (API) (mL)**	**Cake sluggish coefficient**
System 1	112.5	92	20.95	2.5/5	4.8	22	0.141
System 2	102.5	79	24.02	2/3.5	3.8	18	0.061

System 2: 400 mL water + 2% bentonite + 0.3% NaOH
+ 0.5% filtration control agent + 1% silicone viscosity reducer
+ 1% YP increaser + 0.5% amine-inhibitor
+ 4% ultrafine calcium + 1% bonded lubricant + 350 g barite

b. The experimental results of environmentally-friendly drilling fluid systems with density of 1.8 g/cm^3 are shown in Table 2.9

System 1: 400 mL water + 2% bentonite + 0.3% NaOH
+ 0.5% filtration control agent + 1% silicone viscosity reducer
+ 1% polyethylene glycol + 1% YP increaser
+ 0.5% amine-inhibitor + 4% ultrafine calcium + 500 g barite

System 2: 400 mL water + 2% bentonite + 0.3% NaOH
+ 0.5% filtration control agent + 1% silicone viscosity reducer
+ 1% polyethylene glycol + 1% YP increaser
+ 0.5% amine-inhibitor + 4% ultrafine calcium
+ 1% bonded lubricant + 500 g barite

c. The experimental results of environmentally-friendly drilling fluid systems with density of 1.8 g/cm^{3}are shown in Table 2.10.

System 1: 400 mL water + 0.1% potassium + 0.3% PAC-LV
+ 0.2% HEC + 2% starch + 2% ZHFD
+ 1% film-forming agent + 1% bionic inhibitor
+ 1% wellbore strengthener + 2% $CaCO_3$ + 5% KCl
+ 700 g barite

Table 2.10 Environmentally friendly system with a density of 2.0 g/cm^3.

Room temperature	AV (mPa · s)	PV (mPa · s)	YP (Pa)	Gel (Pa/Pa)	FL (API) (mL)	HTHP (API) (mL)	Cake sluggish coefficient
System 1	100	85	15.33	1/6.5			
System 2	97.5	82	15.84	1/4.5			
System 3	97	76	21.46	5/8			
Hot rolling at 120°C for 16 h	**AV (mPa · s)**	**PV (mPa · s)**	**YP (Pa)**	**Gel/ (Pa/Pa)**	**FL (API) (mL)**	**HTHP (API) (mL)**	**Cake sluggish coefficient**
System 1	57	60	−3.07	0/0	8.4	22.4	0.277
System 2	52.5	43	9.71	3.5/4	3.6	10.4	0.109
System 3	51.5	39	12.78	5/6	4.4	12	0.102

System 2: 400 mL water + 0.1% potassium + 0.3% PAC-LV + 0.2% HEC + 2% starch + 2% ZHFD + 1% film-forming agent + 1% bionic inhibitor + 1% wellbore strengthener + 2% $CaCO_3$ + 5% KCl + 1% bonded lubricant + 700 g barite

System 3: 400 mL water + 0.1% potassium + 0.3% PAC-LV + 0.2% HEC + 2% starch + 2% ZHFD + 1% film-forming agent + 1% bionic inhibitor + 1% wellbore strengthener + 2% $CaCO_3$ + 5% KCl + 2% bonded lubricant + 700 g barite

It can be seen from the experimental results that the bonded lubricant has the effects of reducing viscosity and filtration loss in these environmentally friendly systems. The smaller the density of the system, the more obvious the viscosity-reducing effect. The bonded lubricant can significantly reduce the HTHP filtration loss of the system. In addition, the cake sluggish coefficient can also be reduced by about 50%. Overall, the bonded lubricant has good compatibility with the system.

2.2 Mechanism of the bionic bonded lubricants

The purpose of lubrication is to form a lubricating film with a certain normal load-bearing capacity and low tangential shear strength between

Figure 2.9 Adsorption film morphology in the initial stage of adsorption.

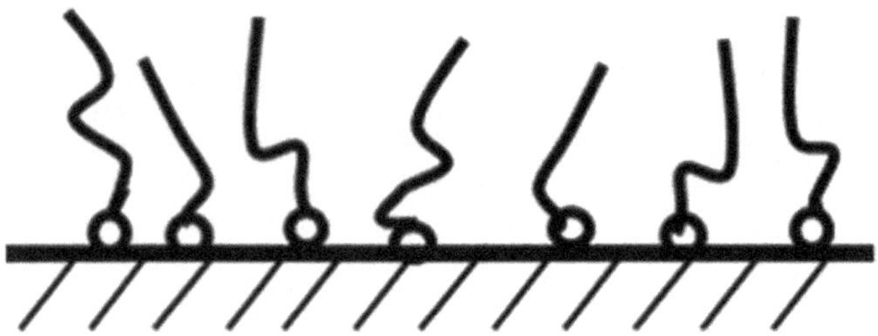

Figure 2.10 The molecular film is compressed into a liquid condensed film.

the friction surfaces. The film is used to reduce material wear and frictional resistance. It can be a fluid film composed of liquid or gas or a solid film. According to the formation principle and characteristics of lubricating films, the lubrication state can be divided into hydrodynamic lubrication, hydrostatic lubrication, elastohydrodynamic lubrication, film lubrication, boundary lubrication, and dry friction states.

In the initial stage of lubrication, in the absence of pressure and relative movement of the friction pair, the polar molecules of the lubricant are adsorbed to the contact surface by the van der Waals force to form a molecular film. At this time, the molecular film pressure is very low and belongs to a gaseous film or a liquid expansion film with good compressibility, and the tail chain has a considerable degree of mobility and can bend or even lie flat (Fig. 2.9).

In the initial stage of lubrication the load is low, and the adsorption film of lubricant molecules is still in the stage of liquid expansion film. As the load increases, the adsorbed molecules are compressed and arranged more densely, and the liquid expansion film begins to transform into a liquid condensed film (Figs. 2.10 and 2.11).

As the load continues to increase, the adsorbed molecules will be further compressed and become denser. The liquid condensed film will begin to transform into a solid film. At this time, the shear strength of the solid film is the main factor determining the friction coefficient (Fig. 2.11).

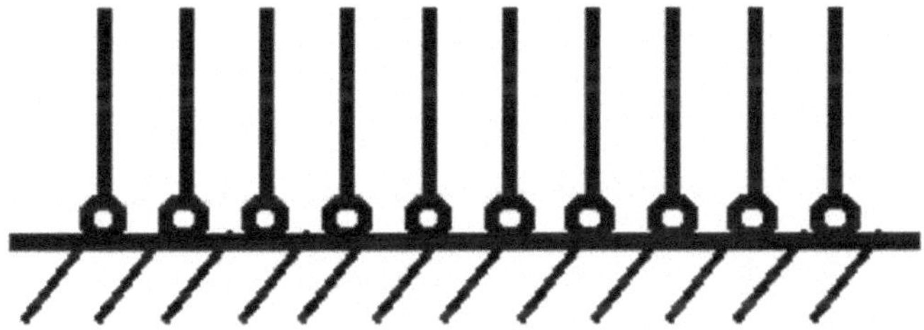

Figure 2.11 The molecular film is further compressed into a solid film.

The polar substances contained in the lubricant are chemically active and can be firmly adsorbed on the metal surface by the van der Waals force between the molecules or atoms, forming an adsorption film featured by layered orientated monomolecular or multimolecular layer [14–16]. This adsorption is called physical adsorption. In addition to individual rough peak points, the adsorption film separates the two friction surfaces, providing an interface with low shear resistance, thus reducing the friction coefficient and avoiding surface sticking. The hydrocarbons with long-chain structure have low physical adsorption capacity, and the formation of physical adsorption film is reversible [17–19]. The physical adsorption film is usually composed of three or four layers of molecules, and each layer of molecules is densely arranged. The cohesion between the molecules gives the molecular grid a certain carrying capacity. Therefore, the two friction surfaces are separated by the adsorption films, and the friction between the films is mainly generated during the sliding of the two surfaces (Fig. 2.12) [20–23].

When the surface temperature is high, polar molecules can form a metal soap with the surface metal and rely on chemical bonding to be adsorbed on the metal surface to form a molecular grid. This adsorption is called chemical adsorption layer [24–26]. The formation of a chemical adsorption film is irreversible, and the film has a low coefficient of friction. Therefore compared with physical adsorption films, chemical adsorption films can work under conditions of higher load, higher speed, and higher temperature layer [27–29].

For the friction pair with a certain sliding speed and load, under the condition of appropriate contact temperature, the components in the lubricant, such as sulfur, phosphorus, chlorine, and other elements in the extreme pressure additive, react with the metal surface to quickly form a thick inorganic film. This chemical reaction film has a high melting point, low shear strength, and a strong connection with the metal surface, which

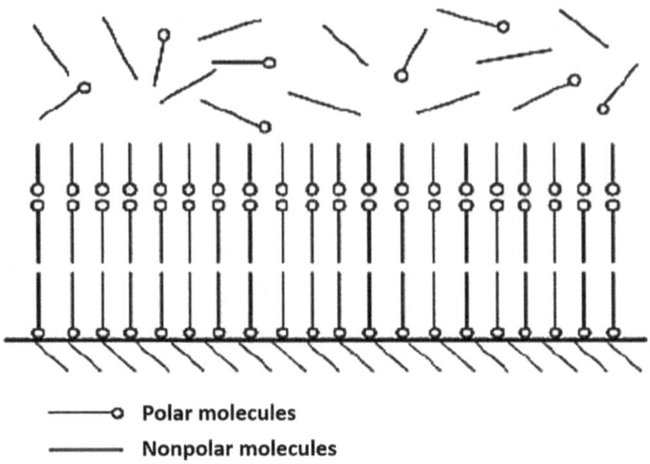

Figure 2.12 Adsorption film model.

can protect the surface from adhesive wear layer [30–32]. In the sliding process, when the reaction film is worn away, a new film will be quickly formed, effectively preventing direct contact between the two friction surfaces.

The chemical reaction film is much more stable than the adsorption film, and its friction coefficient is maintained between 0.10 and 0.25. A good lubrication effect requires a certain thickness for the reaction film, and the thickness of the chemical reaction film is usually 1–10 nm [33]. The formation of the reaction film is irreversible, but during the friction process, the reaction film is constantly worn and generated, so its lubrication effect depends on the dynamic balance of these two processes. If the reaction film fails to generate a new film in time, the lubrication effect will be lost.

The role of the chemical reaction film also depends on its connection strength. Only when the reaction film is firmly connected to the parent metal can it play the role of protecting the metal; otherwise, the reaction film will aggravate the wear. In addition, the chemical properties of the additives that form chemical reaction films are quite lively and easily corrode metals. Therefore the types and amount of additives should be selected according to factors such as friction pair materials and working conditions (Fig. 2.13) [34].

Polar molecules in bonded lubricant carry chemically active polar groups, such as carboxyl and ester groups. These polar groups can be firmly adsorbed on the rock and metal surface by the van der Waals force

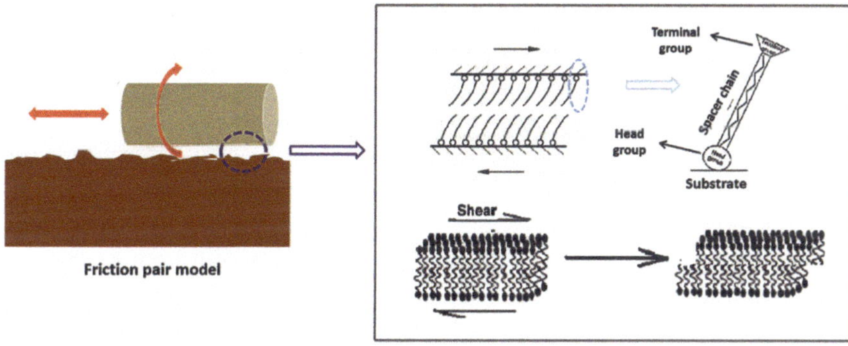

Figure 2.13 Lubrication model and principle.

between molecules or atoms, forming a layered directional monolayer or multilayer physical adsorption film. When the surface temperature is high, the polar molecule forms a metal soap on the metal surface, which relies on chemical bonding to form a chemical adsorption film on the metal surface. The extreme-pressure elements chemically react with the metal surface to quickly form a thick inorganic chemical reaction film. Therefore the solid-solid friction caused by contact between the drilling tool and the borehole rock and between the drilling tool and the casing (or water film contact) will be converted into friction between lubrication films. The polar molecules in the lubricating film are parallel to each other and are perpendicular to the friction surface. This arrangement can satisfy the maximum number of adsorbed molecules. When sliding, under the action of friction, the adsorbed molecules will tilt and bend to form a molecular brush to reduce the resistance, as shown in Fig. 2.12, which greatly reduces the frictional resistance between the drilling tool and the borehole or casing.

Therefore based on the principle of earthworms' reducing viscosity and resistance in soil, a bionic bonded lubricant has been developed, and a series of evaluations have been carried out. The conclusions are summarized as follows:

1. The bionic bonded lubricant has excellent lubricity. Its lubricity can be seen when the dosage is 0.2%, which significantly reduces the lubrication coefficient in 4% base mud. After hot rolling, the lubricity is further improved. When the dosage is 0.4%, the lubrication coefficient in 4% base mud can be reduced by about 90%, which is better than the existing lubrication products.

2. The bonded lubricant can reduce the flow resistance of drilling fluid and can also reduce the adhesion coefficient of cake by more than 50%. The performance of the bonded lubricant in terms of this property is close to that of international advanced lubricants.
3. Compared with the advanced Chinese and international lubricants, the bonded lubricant has superior antiwear performance and can significantly reduce the wear between contact surfaces.
4. The bonded lubricant has good salt and temperature resistance. After a large amount of salt is added, it can maintain good lubrication performance. After hot rolling at 180°C for 16 h, the reduction rate of the lubrication coefficient can still reach 90%.
5. The addition of the bonded lubricant to the polysulfon system can reduce the viscosity and YP of the system, thus improving the rheological properties and reducing the API filtration loss after hot rolling.
6. In a high-density soilless KCl system, the bonded lubricant has obvious effects of viscosity and YP reduction at room temperature. It still has good performance after hot rolling, and the effect of YP reduction is further improved.
7. In an environmentally friendly system, the bonded lubricant has the effect of reducing viscosity and filtration loss. The smaller the density of the system, the more obvious the viscosity-reducing effect, and it can significantly reduce the filtration loss of the system at HTHP conditions. In addition, it can reduce the cake sluggish coefficient by about 50%.
8. The bonded lubricant has no biological toxicity.

References

[1] D. Xiaoqiang, W. Lin, Y. Xiaohua, Advances in the lubricant of water based drilling fluid, Sino-Global Energy 17 (10) (2012) 28–33.
[2] Y. Bin, Y. Wenliang, W. Bin, T. Dehua, Rheological and tribological characteristics of chemically modified castor oil, Lubrication Eng. 1 (2005) 37–38.
[3] L. Guanghuan, L. Tao, T. Zengyan, et al., Research and application of new lubricant synthesized by waste animal/plant oil for drilling fluid, Oilfield Chem. 31 (4) (2014) 488–491. 496.
[4] W. Lin, D. Xiaoqiang, X. Xiaohua, et al., Development and evaluation of a high density drilling fluid lubricant, Drill. Fluid Completion Fluid 33 (1) (2016) 28–32.
[5] L. Huanlong, Basic Research on the Lubrication Characteristics of Water Hydraulic Piston Pump, Southwest Jiaotong University, 2007.
[6] P. Mousavi, W. Dianxia, C.S. Grant, et al., Measuring thermal degradation of a polyol ester lubricant in liquid phase, Ind. Eng. Chem. Res. 44 (15) (2005) 5455–5464.

[7] A.R. Navarro, W.R. Dannels, Maximizing Drilling Operations by Mitigating the Adverse Effects of Friction Through Advanced Drilling Fluid Technology, AADE, Houston, 2011, pp. 1–7.
[8] S. Lei, Research on Low Boiling Point Fuel Lubricity Evaluation Method, Shanghai Jiaotong University, 2012.
[9] W. Weizu, H. Ping, Current Status of Boundary Lubrication Research, South China University of Technology, 2017.
[10] Y. Zhiguang, Practical Handbook of New Lubricating Materials and Lubrication Technology, National Defense Industry Press, Beijing, 1999.
[11] T. Xiaoying, L. Songfeng, W. Shizhu, The effect of liquid crystal on the lubricating properties of mineral oil In: Proceedings of the Fifth National Tribology Conference, Tribology Branch of Chinese Mechanical Engineering Society, National Defense Industry Press, Beijing, 1992, pp. 80–87.
[12] Z. Chunhong, D. Junxiu, W. Shizhu, Study on the matching characteristics between common tribological materials and lubricant anti-friction additives, Tribology 12 (1) (1992) 73–80.
[13] W. Yali, T. Qiuhong, L. Jingyu, Low temperature property of engine lube with compatible environment, J. Changan Univ. (Nat. Sci. Ed.) 3 (2003) 75–78.
[14] Z. Zhijun, Research on the Anti-Wear Properties of Sulfur-Based Additives in Biodegradable Base Oil, Chang'an University, 2006.
[15] C. Nianhuan, N. Maeda, M. Tirrell, et al., Adhesion and friction of polymer surfaces: the effect of chain ends, Macromolecules 38 (8) (2005) 3491–3503.
[16] Z. Hongbo, T. Yu, Z. Boxin, et al., Friction at the liquid/liquid interface of two immiscible polymer films, Langmuir 25 (9) (2009) 4954–4964.
[17] J. Zhiliang, Tribological Properties of Organic Borate in Green Lubricant Base Oil, Changan University, 2007.
[18] P.M. Mcguiggan, M.L. Gee, H. Yoshizawa, et al., Friction studies of polymer lubricated surfaces, Macromolecules 40 (6) (2007) 2126–2133.
[19] A. Blom, C. Drummond, E.J. Wanless, et al., Surfactant boundary lubricant film modified by an amphiphilic diblock copolymer, Langmuir 21 (7) (2005) 2779–2788.
[20] W. Shangyong, An analysis on the leakage in precision pair of high pressure common rail fuel injector, Automot. Eng. 6 (2004) 666–670.
[21] S. Bingshen, Experimental Study and Numerical Simulation of Point Contact Mixed Lubricating Film Thickness, Tsinghua University, 2008, p. 3.
[22] W. Shizhu, H. Ping, Principles of Tribology, 3rd ed., Tsinghua University Press, Beijing, 2008, pp. 200–201.
[23] M. Lei, Z. Fujian, L. Xiaorui, et al., Friction reducers for slick water: performance evaluation and friction prediction, Drill. Fluid Completion Fluid 34 (3) (2017) 105–110.
[24] J. Fengtao, Fluid Dynamics and Finite Element Analysis of Precision Couple Motion Pair, Huazhong University of Science and Technology, 2006.
[25] D. Zhicheng, L. Jun, Exploration of boundary lubrication, Guangdong Sci. Technol. 8 (2006) 41–44.
[26] J. Zhongxuan, Analysis on the factors of sticky sticking, Oil Drill. Prod. Technol. 6 (2) (1984) 37–44.
[27] L. Kuankuan, Study on Modeling and Tribological Properties of Water-Based Lubricants, Beijing Jiaotong University, 2017.
[28] C. Yanbin, L. Jin, Z. Jiangang, et al., Research on the adhesion device of the adhesion coefficient measuring instrument, Equip. Geol. Eng. 18 (1) (2017) 18–20.
[29] W. Weizu, Theoretical and Experimental Research on the Mechanism of Adsorption Films Under Boundary Lubrication Conditions, South China University of Technology, 2006.

[30] M. Huichao, The Preparation of Phosphotungstic (Phosphomolybdic) Acid Ammonium Ultrafine Particles and their Tribological Performance in Water, Henan University, 2013.
[31] F.P. Bowden, D. Tabor, The Friction and Lubrication of Solid, Oxford, 1954.
[32] W. Shizhu, Nano Tribology, Tsinghua University Press, 1998.
[33] S. Huaiwen, Research on Elastic Hydrodynamic Lubrication With Elliptical Contact Pure Extrusion, Qingdao Technological University, 2007.
[34] L. Kaihe, A summary of environmentally-friendly drilling fluid lubricants, Oil-Gasfield Surf. Eng. 7 (2004) 29.

CHAPTER 3

New technology of bionic drilling fluids for intelligent leakage prevention and plugging

Lost circulation is a complex type of downhole situation in which various working fluids (including drilling fluids, cement slurries, completion fluids, and other fluids) enter the formation under the action of a pressure differential during downhole operations such as drilling, well cementing, and testing. Lost circulation is one of the most common technical problems in drilling engineering. It not only consumes much drilling time, wastes drilling fluids, damages formations, and interferes with geological logging, but also can cause a series of complex conditions and accidents such as well collapse, pipe sticking, and well blowout and can even cause the wellbore to be scrapped, leading to significant economic losses [1–17]. Therefore plugging materials are indispensable in the process of leakage prevention and plugging. The success rate of leakage prevention and plugging depends directly on the performance of the plugging materials. According to their mechanisms and functions, existing plugging materials can be divided into bridge plugging materials, high-dehydration plugging materials, temporary plugging materials for protecting formations, chemical plugging materials, inorganic gel plugging materials and soft (hard) plug plugging agents, composite plugging materials, and general-purpose temporary plugging materials for protecting formations [18–22]. Although these materials and technologies have achieved some success, they still face some challenges. For example, it is difficult to accurately predict the position of leakage zone, the pressure of leakage zone is uncertain, the opening size of leakage channels cannot be determined; the success rate of one-time plugging is low, especially for vicious leakage; it is difficult to form a stable pressure-bearing plugged zone on a smooth crack wall surface, and the bearing pressure is low; the use of solid particles with different particle sizes might not match the leakage pores and cracks well enough, leading to a poor plugging effect; the plugged zone formed by plugging materials is easily diluted by the formation fluid (such as water),

Fundamentals and Applications of Bionic Drilling Fluids
DOI: https://doi.org/10.1016/B978-0-323-90293-9.00003-4

leading to plugging failure; and plugging materials may damage formations. Furthermore, no uniform experimental device and no evaluation standard for leak prevention and plugging currently exist. How to conduct a comprehensive and accurate evaluation of leakage prevention and plugging agents and plugging effect indoors is a technical problem that needs to be solved urgently, thereby providing experimental data for on-site leakage prevention and plugging technology [23–33].

To this end, Professor Guancheng Jiang of China University of Petroleum (Beijing) led his team to introduce bionics into the field of drilling fluids. By simulating the high-strength, high-toughness, uniform-spatial-force vein structure in royal water lily leaves, which is formed by the connection of multiple nodes, and combining the simulation with the multilayer composite structure of shells, a new material has been developed for plugging under stop and pressure conditions. The material consists of deformable fiber mesh, high-strength gel, and rigid particles. By imitating the principle of the automatic hemostasis of mammalian blood, a new material has been developed for plugging under in-drilling conditions. The plugging material is composed of nanowhiskers, weak gels, and particles. On this basis an efficient and safe new technology for plugging under in-drilling and stop conditions has been formed, and good results have been achieved in laboratory and field tests.

3.1 Bionic plugging technology for stop conditions based on the buoyancy of royal water lily leaves and the strong bearing pressure of the composite structure of shells

3.1.1 Research on leakage prevention and plugging technology of bionic drilling fluid based on the structure of royal water lily leaves

Royal water lily leaves have great buoyancy because the backs of the leaves have thick veins similar to spider webs. The center of the back of the leaf is the petiole, and from the petiole, numerous thick and hollow veins are arranged radially. Between the large veins there are thin veins with a sickle shape. There are also air chambers inside the veins, forming interlaced skeleton structure, as shown in Fig. 3.1. It is this structure that give the leaves of royal water lily a large pressure-bearing capacity and allows them to float steadily on water [34].

During the drilling process, almost all places with lost circulation have pores and cracks. Under the effect of a positive pressure differential, the

Figure 3.1 Royal water lily vein structure.

drilling fluid penetrates and flows into the formation through the pores and cracks, leading to complex downhole conditions and even leakage accidents. This can bring great economic losses to enterprises and even cause drilling failure. The best way to avoid this situation is to add plugging materials to the drilling fluid system to plug the leakage zone so that the drilling fluid will not flow through the pores and cracks to the inside of the formation under a positive pressure differential and increase the pressure-bearing capacity of the borehole.

Royal water lily leaves float on the surface of water and have great pressure-bearing capacity. Inspired by their prodigious buoyancy, Professor Guancheng Jiang of China University of Petroleum (Beijing) believes that the materials with the following characteristics can avoid the leakage of drilling fluids during the drilling process. (1) The vein structure forms a spider web at the pores and cracks. (2) Particles similar to petioles fill the inner space of the spider web. (3) The material with the shape of a barbed sickle further plugs the space between the spider web and the particles, and enhances the cohesion between the spider web, the particles, and the formation rock to increase the bearing pressure. Therefore he led his team members to successfully develop a bionic drilling fluid plugging material composed of fiber material with a leaf vein structure, solid microparticle material with a petiole function, and gel material connecting the simulated leaf veins. This material enabled the new technology of bionic drilling fluid for plugging under stop conditions.

3.1.2 Research on leakage prevention and plugging technology of bionic drilling fluid based on the multilayer composite structure of shells

Nature provides highly sophisticated mechanisms for the design and synthesis of materials. Materials produced by living organisms often have better physical properties than those synthesized by similar ingredients. Compared with many artificial materials formed from top to bottom, the nanocomposites formed by using bottom-up self-assembly methods in nature have higher strength and better toughness, and shells are the best example. Shells are composed of about 95% inorganic phase $CaCO_3$ (calcite and aragonite) and a few percent of organic biopolymers. Although their composition is brittle, they show good comprehensive mechanical properties [35], especially a high level of toughness. Therefore the toughening mechanism of shells is of great significance for material design and preparation. According to the structure of shells and the crack morphology and propagation mode, it can be considered that there are four main toughening mechanisms in shells: fiber pull-out, crack deflection, organic matrix bridging, and mineral bridge.

First, fiber pull-out can absorb more energy and toughen the material, which is an important mechanism for toughening fiber-reinforced composite materials. The binding force and friction between the organic phase and the aragonite layer will prevent the crystals from being pulled out. At the same time, the microconvex body under the surface of the aragonite plate has a locking effect on the mutual movement of adjacent plates, thereby increasing the resistance of crack propagation and fiber pull-out and increasing the material's toughness.

Second, when deformation and fracture occur in the shell, the organic matrix between the inorganic phases ($CaCO_3$) undergoes plastic deformation and adheres well to the adjacent inorganic phases. This is a common phenomenon in shells, indicating that the biomacromolecules and inorganic phases have a strong binding interface, which improves the sliding resistance between adjacent crystals and strengthens the role of the fiber pull-out's toughening mechanism. In addition, after plastic deformation the organic phase still maintains a good combination with the inorganic phase, which play a bridging role between the separated inorganic phases, thereby reducing the stress at the crack tip, increasing the resistance of crack propagation, and improving the toughness. This toughening mechanism is called organic matrix bridging [36].

It has been found that the total area of the mineral bridge in the nacre accounts for about one-sixth of the total area of the aragonite plate, and its influence on the overall mechanical properties of the nacre cannot be ignored. During fracture of the nacre, owing to the existence of the mineral bridge and the randomness of its position, the resistance to crack propagation and the effect of crack deflection are increased, and a mineral bridge often remains between the upper and lower aragonite plates that have fractured. Therefore to pull out the aragonite crystals, in addition to overcoming the binding force between the organic phase and the aragonite tablet, the mineral bridge between the aragonite crystals must be cut so that the toughness of the nacre can be strengthened [37].

These toughening mechanisms are closely related to the special composition and structure of shells. The reason is that organisms use their own powerful control over the construction of materials, through the modulation of cells, to regularly embed biopolymers in inorganic minerals or between mineral crystals. Although the proportion of organic matter is small, it can fundamentally change the fracture characteristics of the material, thereby greatly improving its toughness [38]. At present, the commonly used drilling fluid leakage prevention and plugging materials have the disadvantages of weak adhesion to the smooth wall, low bearing pressure of the plugged zone, and the tendency of the plugged zone to be diluted by formation fluids. The process and mechanism of forming the composition and structure of shells undoubtedly have important implications for the design and preparation of leakage prevention and plugging drilling fluid materials [37]. At present, the team led by Guancheng Jiang of China University of Petroleum (Beijing) has successfully developed and widely applied shell-like drilling fluid materials, and significant achievements have been made. However, to imitate the biomineralization mechanism of shells, the preparation of microlayered composite materials through self-assembly needs to be further strengthened. It will be more and more valued in the design and preparation of future high-performance drilling fluid materials and will play an important role.

3.1.3 Establishment and performance evaluation of the new technology of bionic drilling fluid for plugging under stop conditions

By simulating royal water lily leaves connected by multiple nodes to form a high-strength, high-toughness, uniform-spatial-force vein structure, combined with the multilayer composite structure of shells, a new material has been developed for plugging under stop and pressure conditions. The material consists of deformable fiber mesh, high-strength gel, and rigid particles. The deformable fiber mesh is used to imitate the vein structure of royal water lily leaves; the composite of high-strength gel and rigid particles imitates the multilayer structure of the shell consisting of the outermost cuticle, the middle prism layer, and the bottom pearl layer. The composition of shells contains more than 95% inorganic substances and less than 5% organic substances. The process of shell mineralization is simulated, along with the process of self-assembly by chemical recognition growth, hydrogen bonding, and other chemical bond layers, thereby achieving the compound effects of structural bionics, chemical composition bionics, process bionics, and mechanism bionics. Specifically, the high-strength gel is a supramolecular gel plugging material developed on the basis of the theory of supramolecular chemistry and is prepared by introducing special functional groups on the macromolecular chain. In aqueous solution the functional groups of the macromolecular chain spontaneously form an ordered supramolecular structure through noncovalent bond interaction, which is reversible and intelligent. During the drilling process, when leakage occurs, the bionic plugging material plays its role through the effects of fiber pull-out, organic matrix bridging, and mineral bridge to achieve successful plugging.

3.1.3.1 Establishment and performance evaluation of supramolecular bionic plugging technology for stop conditions suitable for medium- and high-temperature formations

A gel-type plugging agent for stop conditions has obvious advantages in plugging for high-temperature formations. Therefore based on the principles of bionics and drilling fluid technology, a supramolecular plugging agent for stop conditions suitable for medium- and high-temperature formations has been developed, and a medium- and high-temperature supramolecular bionic drilling fluid technology for plugging under stop conditions has been established. Systematic performance evaluation has been conducted, including shear thinning behavior, plugging performance, water washout resistance, and strength evaluation. At the same time, for formations with serious lost circulation, the gel plugging agent can be mixed with other plugging materials to enhance its strength for application in formations with serious lost circulation.

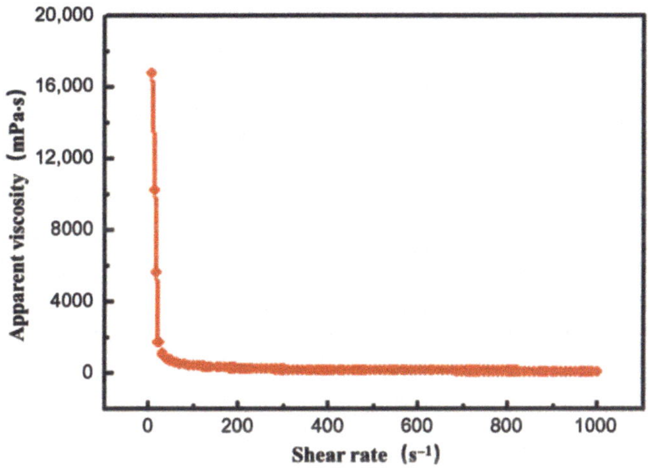

Figure 3.2 Shear-thinning behavior curve of the new gel plugging agent.

3.1.3.1.1 Evaluation of shear-thinning behavior

A HAAKE RS6000 (German) rheometer was used to examine the shear-thinning behavior of the gel. The results are shown in Fig. 3.2. For the gel plugging agent the better the shear-thinning behavior, the better the pumping and plugging, because when the gel was pumped, the gel viscosity was low, and the flow resistance was small, leading to easy flow; when the agent entered the formation, the shear rate decreased, the gel viscosity increased, the flow resistance increased, and finally the flow stopped, gradually forming a gel and plugging the leakage zone. The experimental results show that the gel plugging agent has good shear-thinning behavior. At $1000\ s^{-1}$ the viscosity was $80\ mPa \cdot s$, which was a small value and conducive to pumping; at $5\ s^{-1}$ the viscosity was $16{,}800\ mPa \cdot s$, which was a large value and conducive to the formation of gel and the plugging of the leakage zone.

3.1.3.1.2 Evaluation of plugging performance

The formation's leakage channels were simulated by quartz sand particles of different sizes. The mesh sizes of quartz sand selected in the experiment were 10–20, 20–40, 40–70, and 70–120.

1. Experimental steps

   The plugging performance of the new gel plugging agent was evaluated by a 71-type HTHP water loss meter, and different leakage conditions of the formation were simulated by adjusting the quartz sand mesh. The main experimental steps were as follows:

a. Configure the gel base fluid according to the formula of the new gel plugging agent, and weigh 100 g of quartz sand with the specified mesh number.
b. Place the sealing ring at the bottom of the mud cup of the 71-type HTHP water loss meter, and then pour 100 g of quartz sand and 350 mL of gel base liquid into the mud cup at a time and put on the top cover.
c. Place the mud cup on the base of the 71-type HTHP water loss meter, set the experimental temperature, and install the pressurizing device.
d. Place a measuring cylinder at the lower valve stem to record the filtration loss of the sand bed. After the temperature rises to the specified temperature, place the cylinder at this temperature for a certain period of time to simulate the gel-forming process.
e. After the gel forms, unscrew the valve stem from the upper air inlet of the HTHP water loss meter to make the pressure enter the steel cylinder of the HTHP water loss meter, gradually pressurize from 0.7 to 3.5 MPa, and maintain each pressure for 30 min.
f. Observe the pace of filtration loss in the measuring cylinder. and record the values.

2. Experimental results

The test results at 90°C—120°C are shown in Tables 3.1—3.4. As can be seen, the new gel plugging agent can withstand more than 3.5 MPa at different temperatures, and the change trend of filtration loss with quartz sand particle size is approximately the same as that with temperature. At the same temperature the filtration loss decreased as the quartz sand particle size decreases; at the same quartz sand particle size, the filtration loss decreased as the temperature increases. This is because as the gelation time decreased with increasing temperature, the leached liquid during nongelatinization process further decreased, and at the same time, the strength of the formed gel increased, so the filtration loss decreased. In addition, under the same conditions (such as 10—20 mesh sand bed at 3.5 MPa), the filtration loss of the new gel plugging agent was less than 10% of the gel base fluid, showing a good plugging effect.

3. Evaluation of plugging capacity after compounding

Amounts of 5%, 10%, and 15% of ultrafine calcium carbonate were added to the basic formula of the new gel plugging agent for compounding, and the plugging ability after compounding was evaluated.

Table 3.1 Cumulative filtration loss for quartz sand with different particle sizes at 90°C.

Pressure (MPa)	Quartz sand particle size (mesh)			
	10–20	20–40	40–70	70–120
0.7	15 mL	12 mL	10 mL	7 mL
1.5	19 mL	15 mL	11 mL	8 mL
2	21 mL	19 mL	12 mL	9 mL
2.5	22 mL	21 mL	13 mL	9 mL
3	24 mL	21 mL	13 mL	9 mL
3.5	24 mL	21 mL	13 mL	9 mL

Table 3.2 Cumulative filtration loss for quartz sand with different particle sizes at 100°C.

Pressure (MPa)	Quartz sand particle size (mesh)			
	10–20	20–40	40–70	70–120
0.7	12 mL	11 mL	8 mL	6 mL
1.5	13 mL	13 mL	10 mL	7 mL
2	15 mL	14 mL	11 mL	8 mL
2.5	16 mL	15 mL	12 mL	8 mL
3	18 mL	15 mL	12 mL	8 mL
3.5	18 mL	15 mL	12 mL	8 mL

Table 3.3 Cumulative filtration loss for quartz sand with different particle sizes at 110°C.

Pressure (MPa)	Quartz sand particle size (mesh)			
	10–20	20–40	40–70	70–120
0.7	10 mL	8 mL	7 mL	5 mL
1.5	12 mL	10 mL	8 mL	6 mL
2	13 mL	11 mL	9 mL	7 mL
2.5	14 mL	12 mL	10 mL	7 mL
3	15 mL	12 mL	10 mL	7 mL
3.5	15 mL	12 mL	10 mL	7 mL

Fig. 3.3 shows, from left to right, the gel formed after mixing with 0%, 5%, 10%, and 15% concentrations of ultrafine calcium carbonate. The experimental data of 120°C and 40–70 mesh are shown in Table 3.5. Under the same pressure, the cumulative filtration loss decreased with the

Table 3.4 Cumulative filtration loss for quartz sand with different particle sizes at 120°C.

Pressure (MPa)	Quartz sand particle size (mesh)			
	10–20	20–40	40–70	70–120
0.7	8 mL	7 mL	6 mL	3 mL
1.5	10 mL	8 mL	7 mL	4 mL
2	11 mL	10 mL	9 mL	5 mL
2.5	12 mL	11 mL	10 mL	6 mL
3	13 mL	11 mL	10 mL	6 mL
3.5	13 mL	11 mL	10 mL	6 mL

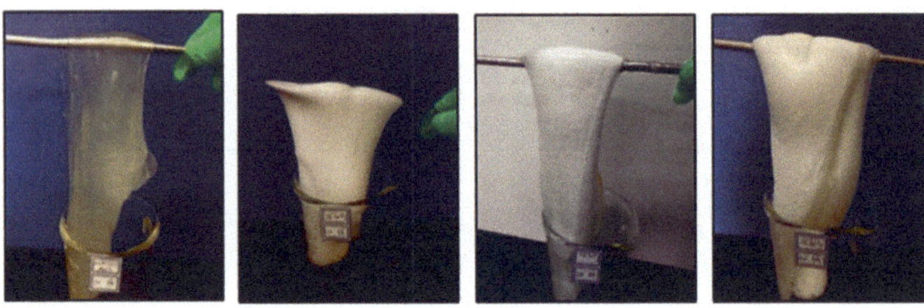

Figure 3.3 Images of gelation of ultrafine calcium carbonate with different concentrations.

Table 3.5 Cumulative filtration loss of ultrafine calcium carbonate with different concentrations.

Pressure (MPa)	Concentration of ultrafine calcium carbonate			
	0%	5%	10%	15%
0.7	6 mL	6 mL	4 mL	2 mL
1.5	7 mL	7 mL	5 mL	3 mL
2	9 mL	8 mL	6 mL	3 mL
2.5	10 mL	9 mL	6 mL	3 mL
3	10 mL	9 mL	6 mL	3 mL
3.5	10 mL	9 mL	6 mL	3 mL

increase in ultrafine calcium carbonate concentration; under the same ultrafine calcium carbonate concentration, as the pressure increased, the curve became flat, that is, the cumulative filtration loss remained at a certain value and no longer changed because with the occurrence of filtration

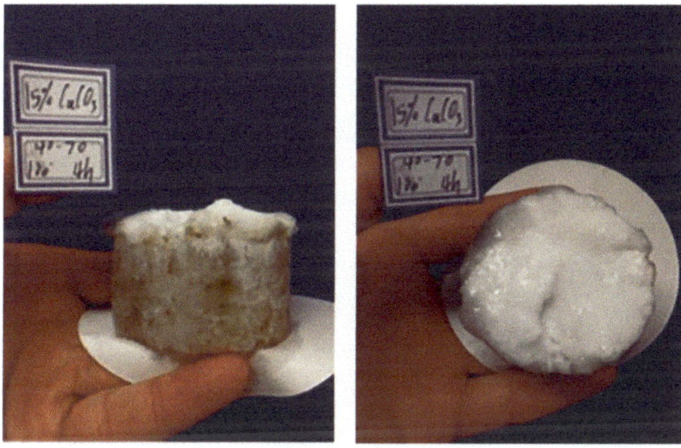

Figure 3.4 Side and top views of the sand bed after plugging.

loss, a dense white filter cake gradually formed on the surface of the sand bed, preventing the occurrence of filtration loss (Fig. 3.4). The higher the concentration of ultrafine calcium carbonate, the denser the filter cake and the earlier the flat curve appeared.

3.1.3.1.3 Resistance to water washout

After the gel base fluid entered the formation, a gel was formed at a certain temperature, and then a plugged zone was formed, but when the gel reacted with the formation water or was miscible, the plugged zone did not exist and the plugging failed. So the gel's resistance to water washout has a very important effect on the plugging effect. The following content will analyze the water washout resistance of the new gel plugging agent.

1. Experimental method
   a. Prepare the formation water according to the formula of 400 g water + 28 g NaCl + 2.4 g $CaCl_2$ + 1.6 g $MgCl_2$, pour it into a 600-mL beaker, and maintain the preconfigured gel base fluid at a certain temperature to form a gel.
   b. Place the gel in a beaker filled with formation water, cover it with plastic wrap, and place it in a dry box at a certain temperature to simulate the formation environment.
   c. Manually stir the gel in the formation water every 24 h, and observe the state between the gel and the formation water.

2. Experimental results

It can be seen from Fig. 3.5 that after being mixed with the formation water, the gel was not miscible with the water within 3 days, with only slight swelling. It was still not miscible or diluted by manual stirring, with very strong pick-hanging performance, indicating that the new gel plugging agent has strong resistance to water washout.

3.1.3.1.4 Evaluation of gel strength

The cone-plate system of the HAAKE RS6000 rheometer was used, and the inflection point method of the flow curve in the controlled shear rate mode in the steady state test, that is, the CR mode method, was used to

Figure 3.5 The mixed state of the gel and formation water within 72 h.

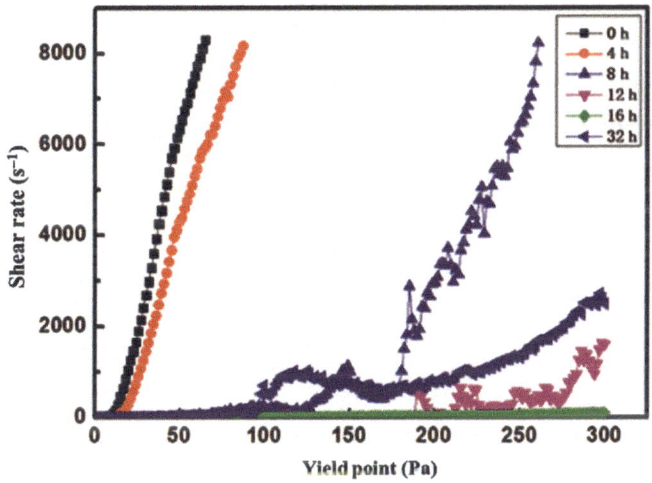

Figure 3.6 CR mode test curves with different gelation times.

test the yield stress so as to examine the strength of the gel formed by the new plugging agent under a different holding time at 120°C.

The experimental results are shown in Fig. 3.6. During the holding time of 0–4 h, the yield stress corresponding to the curve inflection point was 20 Pa. When the gelation time continued to increase, the yield stress gradually increased. When it reached 16 h, the yield stress exceeded 300 Pa. This was because as the gelation time increased, the interior grid structure of the gel gradually increased and the gel strength gradually increased. When the gelation time continued to increase to 32 h, the yield stress corresponding to the curve inflection point was 200 Pa, which showed that although the long-term holding time at high temperature would damage the grid structure, the generation rate of the grid structure was greater than the breaking rate under the action of temperature, so the new gel plugging agent still had high strength at high temperature and could maintain the stability of the structure for a long time.

3.1.3.1.5 Evaluation of unplugging performance

If the new gel plugging agent is used for plugging formations, it must meet the requirements of plugging and unplugging at the same time; that is, after plugging is completed, it must be able to flow back smoothly and have the capability of unplugging. Here, an oxidizing agent (gel breaker) was used to evaluate its plugging performance.

Table 3.6 Gel breaking rate of gel breakers with different concentrations at 90°C–120°C within 72 h.

Gel breaking rate (%)	Concentration of gel breaker solution (%)			
Temperature (°C)	**5**	**10**	**15**	**20**
90	60.7	73.8	95.5	95.4
100	64.5	78.6	98.6	98.9
110	66.8	84.8	98.6	98.9
120	71.0	89.6	99.2	99.2

The gel-breaking rate of the gel breakers with different concentration within 72 h at a temperature range of 90°C–120°C is shown in Table 3.6. It can be seen that at any temperature the breaking efficiency of the gel breakers with 15% and 20% concentrations in the same period is the highest. Considering the gel-breaking rate and drilling cost, the optimal concentration for gel breaker is 15%, and the gel-breaking rate can reach more than 95%.

3.1.3.2 Establishment and performance evaluation of high-strength acid-soluble gel plugging technology for stop conditions suitable for low- and medium-temperature formations

According to the above research and development ideas, a high-strength acid-soluble plugging agent for stop conditions that is suitable for low and medium temperatures was developed, and the corresponding plugging technology was then established. In this research, the steel ball bed was used to determine the plugging pressure of the gel, and then the method of adjusting the gelation time in different temperature ranges was studied.

3.1.3.2.1 Pressure-bearing evaluation experiment of steel ball bed plugging

Two layers of 8-mm steel balls were added to the mud cup of the HTHP water loss meter to form a steel ball bed. Three samples of 50 mL of plugging slurry were prepared with formulas 1, 2, and 3 and poured into the mud cup to be heated at 40°C and then left for 5 h to cure. Then the mud cup was taken out and opened for observation of the curing of the plugging slurry. Then 100 mL of pure water was added, and the mud cup was put back. The mud cup was pressurized with nitrogen and maintained for 10 min at each pressure. The ability of the gel plugging material to plug the steel ball bed after the addition of different proportions of

Figure 3.7 Cured steel ball bed by formula 4.

Figure 3.8 Cured steel ball bed by formula 5.

Figure 3.9 Cured steel ball bed by formula 6.

heavy calcium carbonate was evaluated through measuring the loss of pure water. After the experiment was completed, the cured steel ball bed was taken out, as shown in Figs. 3.7–3.9. The cured steel ball bed had a complete appearance and a certain amount of strength. At the same time,

Table 3.7 Experimental results of lost circulation rate under different pressures.

Pressure (MPa)	Lost circulation rate (mL/s)		
	Formula 4	Formula 5	Formula 6
1	0	0	0
2	0	0	0
3	0.2	0.012	0.04
4	Completely lost	Completely lost	0.125
5	–	–	Completely lost

the lost circulation rates of formulas 4–6 under different pressure differentials were measured, as shown in Table 3.7. With the increase in the proportion of 3000-mesh heavy calcium carbonate in the gel plugging material, the capacity of the gel plugging material to plug the two-layer 8-mm steel ball bed significantly improved at 40°C, indicating that the addition of heavy calcium carbonate enhanced the compactness of the plugging slurry after curing, thereby enhancing its strength. When the proportion of 3000-mesh heavy calcium carbonate was 20%, the plugging slurry prepared by formula 6 could withstand a pressure of 4 MPa, so the proportion of 20% was used as the basis for the following optimization of the gel material formula.

3.1.3.2.2 Adjustment of gelation time

The gelation time of the AGS ternary gel system is different at different temperatures, so it is necessary to study its gelation time.

1. Method of adjusting gelation time at 40°C–70°C
   - a. Retarder method

     On the basis of the basic formula the proportion of retarder was increased to delay the curing process of the plugging slurry and extend the gelation time. The experimental results are shown in Table 3.8. The gelation time can be extended by adding retarder. When the dosage of retarder was 4%, the gelation time was extended to about 5 h.
   - b. Method of reducing the concentration of the main agent

   Because reducing the AGS concentration can slow down the reaction rate and thus extend the gelation time, the method of reducing the concentration of the main agent was used to prolong the gelation time. The experimental results are shown in Table 3.9. As can be observed, reducing the concentration of the main agent AGS effectively extended the gelation time. However, if the concentration of AGS was too low, the gel strength after gelation was too low to meet the needs of plugging. Therefore it is necessary to synergistically consider the AGS concentration and the dosage of retarder.

2. Optimization of gelation time at 40°C–70°C

In the on-site plugging construction, to improve the plugging effect and reduce the cost, the plugging agents are usually used in combination with inorganic inert materials. For this reason the viscosity and YP of the supramolecular gel plugging agent before gelation need to be considered in order to achieve the purpose of suspending and carrying inorganic inert materials. In the research process, the appropriate amounts of retarder and AGS were chosen to prepare formulas 7, 8, 9, and 10. The experiment was conducted at the corresponding temperatures. The results are shown in Table 3.10 and Fig. 3.10.

The results show that the fluidity of formula 10 became worse, while formulas 7, 8, and 9 did not encounter this situation. To further extend the gelation time of the gel system, the dosage of retarder was further

Table 3.8 Gelation time after adding different dosages of retarder at 40°C.

Temperature (°C)	Dosage of retarder			
	0%	2%	3%	4%
40	1 h	2 h	4 h	5 h

Table 3.9 Results of gelation time after adjusting the AGS concentration.

Temperature (°C)	AGS concentration (%)		
	10%	15%	20%
70	5.5 h	3.5 h	1.2 h

Table 3.10 Gelation time of each formula at different temperatures.

Temperature (°C)	Formula	Gelation time
40	7	3 h 10 min
50	8	3 h 10 min
60	9	2 h 30 min
70	10	4 h 30 min

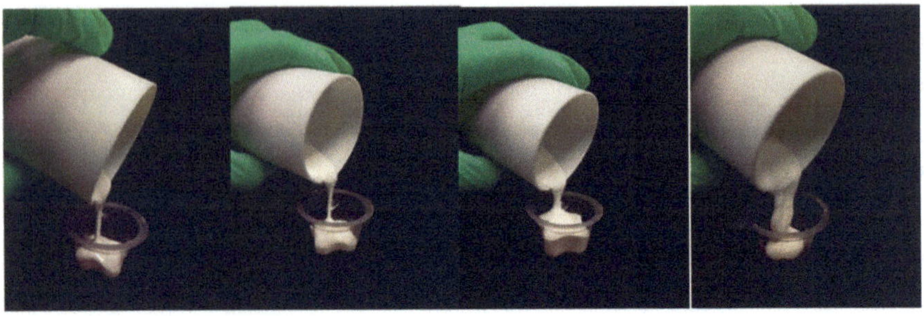

Figure 3.10 Flow state of each gel system at different temperatures (from left to right: formulas 7, 8, 9, and 10).

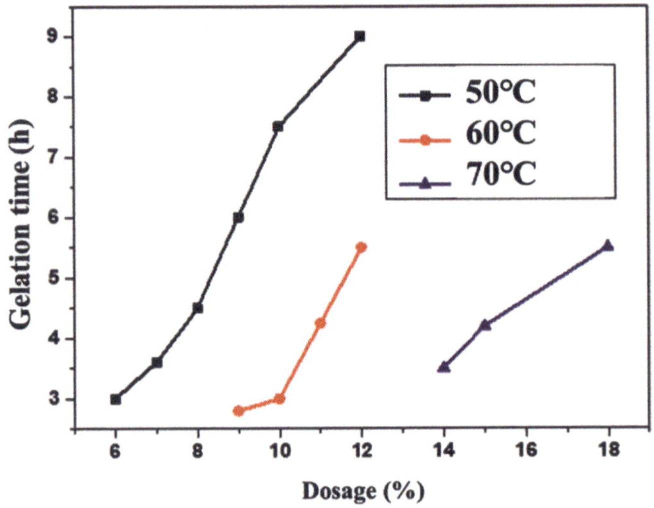

Figure 3.11 The gelation time after further adjusting the dosage of retarder at 50°C–70°C.

increased, and the gelation time performance was investigated. The experimental results are shown in Fig. 3.11. At 50°C, when the dosage of retarder was increased to 9%, the gelation time was extended to 6 h. At 60°C, when the dosage of retarder was 12%, the gelation time was extended to 5.5 h. At 70°C a dosage of retarder of 18% was needed to extend the gelation time to 5.5 h. As a result, a gel plugging formula with a gelation time of about 5.5 h at 40°C–70°C was formed, as shown in Table 3.11.

In addition, this plugging technology can be mixed with the drilling fluid system. For example, 20% and 25% of the plugging slurry were

Table 3.11 High-strength acid-soluble gel plugging formulae for stop conditions at 40°C–70°C.

T (°C)	Formula
40	Pure water + 20% main agent AGS + 0.2% polymer + 20% 3000-mesh calcium carbonate + 5% retarder
50	Pure water + 20% main agent AGS + 0.2% polymer + 20% 3000-mesh calcium carbonate + 9% retarder
60	Pure water + 20% main agent AGS + 0.2% polymer + 20% 3000-mesh calcium carbonate + 12% retarder
70	Pure water + 20% main agent AGS + 0.2% polymer + 20% 3000-mesh calcium carbonate + 18% retarder

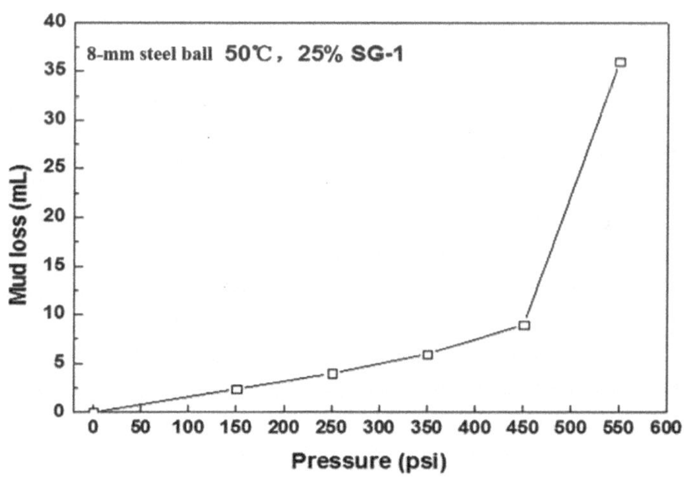

Figure 3.12 Evaluation of the plugging effect under stop conditions at 50°C.

added to the following drilling fluid system (formula 1) to evaluate the plugging effect:

1: 3% bentonite mud + 0.2% NaOH + 0.2% macromolecular emulsion + 3% asphalt + 2% KFT + 0.4% PAC-LV + 2% $CaCO_3$ + 5% KCL + barite (density = 1.4 g/cm^3)

2: Formula 1 + 20% AGS + 0.2% polymer

3: Formula 1 + 25% AGS + 0.2% polymer

According to Fig. 3.12, the plugging slurry could withstand a pressure of 3 MPa at 50°C, and the gelation time was controlled at 4 h. After the

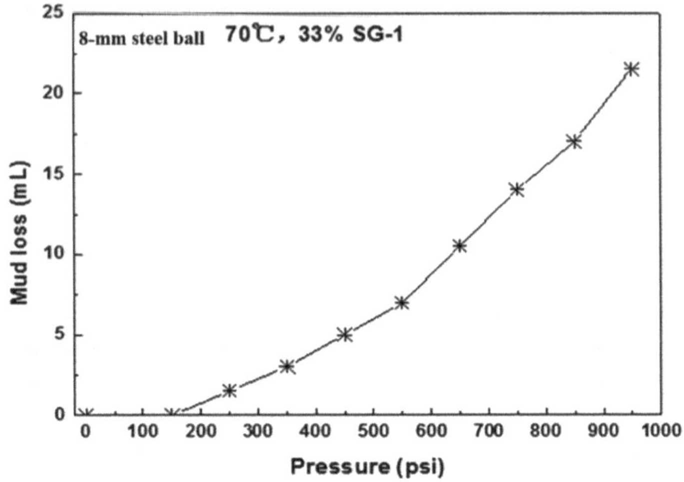

Figure 3.13 Evaluation of the plugging effect under stop conditions at 70°C.

gel plugging agent was added, the AV of the drilling fluid increased slightly, but it did not affect the pumpability before gelation. Fig. 3.13 indicates that the plugging slurry could withstand pressure above 6.5 MPa at 70°C.

3.2 Bionic plugging technology for in-drilling conditions based on the principle of the automatic hemostasis of mammals

The principle of mammals' automatic hemostasis is that platelets contain a series of factors related to blood coagulation, that is, platelet factor 3(PF3), also known as platelet phospholipid. After the platelets adhere and aggregate, biologically active substances (5-TH) are released, which promote the formation and release of plasma activators, activate the fibrin reaction system, and convert the fibrin source in plasma into insoluble blood fibers. The blood fibers are then interwoven into a net that includes many blood cells and forms blood clots, so after the blood leaves the blood vessel for a few minutes, it changes from a flowing sol to a nonflowing jelly clot, that is, automatic hemostatic. This process is actually a series of limited hydrolysis of proteins. Once the coagulation process begins, each coagulation factor forms part of a waterfall-like reaction chain until the blood coagulates [39].

Figure 3.14 The plugging principle of traditional plugging materials for in-drilling conditions, that is, by the bridging and filling of inert materials.

Inspired by this, the team led by Professor Guancheng Jiang of China University of Petroleum (Beijing) has invented a new material for plugging under in-drilling conditions, in which the plugging material is composed of nanowhiskers, weak gels, and particles. Microparticles simulated the platelets, nanowhiskers, and weak gel to simulate fibrinogen so as to use the effects of gel deformation, fiber pull-out, organic matrix bridging, and particulate bridges to automatically achieve plugging when lost circulation is encountered during drilling. When no lost circulation occurs, the new material helps to form dense filter cakes on the borehole and reduce filtration loss. From the theoretical analysis the bionic plugging material developed by using the principle of mammals' automatic hemostasis can accomplish intelligent plugging under in-drilling conditions, avoiding the disadvantage in which inert materials and the principle of bridging and filling are used for plugging under in-drilling conditions, and the size of plugging material is required to be strictly matched with that of leakage pores and cracks (see Figs. 3.14 and 3.15), because the size of leakage pores and cracks cannot be accurately measured before drilling. According to these ideas, we have developed a strong-adhesion expandable gel-fiber composite plugging agent for in-drilling conditions and a supramolecular microgel plugging agent for in-drilling conditions. Then the bionic plugging technologies for in-drilling conditions corresponding to the above two agents were established.

3.2.1 Research and development and performance evaluation of strong-adhesion expandable gel-fiber composite plugging technology for in-drilling conditions

Based on bionics and supramolecular chemistry theory, a strong-adhesion water-swellable cross-linked gel was developed. During the synthesis of

Figure 3.15 The plugging principle of bionic plugging materials for in-drilling conditions, that is, by intelligent plugging under in-drilling conditions.

the gel, dispersible fibers were introduced to form a cage-like structure to further increase the rigidity and strength of the gel to satisfy the pressure-bearing capacity requirements and to achieve effective plugging under in-drilling conditions. The unique three-dimensional (3D) network structure of the cross-linked polymer made it easy for water to enter the structure, leading to gel swelling and good elasticity. The interpenetrating network structure was formed by the composite high-adhesion gel to further improve the cross-linking strength of the gel and its adhesion. The strong-adhesion gel itself could also be adsorbed in the wall of the leakage channels to achieve retention plugging and improve the success rate of plugging. At the same time, biomacromolecules with strong adhesion were added as a matrix in the polymerization process to form an interpenetrating network gel with the 3D network structure gel to improve its gel strength and adhesion. According to the above method, different strong-adhesion expandable gel-fiber composite plugging materials for in-drilling conditions were synthesized, as shown in Fig. 3.16.

3.2.1.1 The plugging effects of a high-temperature-resistant microgel plugging agent SAP-1 and fiber compounding

Different compounding ratios of the microgel plugging agent SAP-1 and fiber were added to the drilling fluid to conduct plugging experiments. The formulas are as follows:

1: (drilling fluid system): Water + 0.3% NaOH + 2.5% PF-FLOTROL + 5% PF-Greenseal + 5% PF-Ezcarb + 0.5% PAC-LV + 4% PF-Soltex + 0.3% PF-Plus + 16% KCl

2: Formula 1 + 1.5% SAP-1 (coarse) + 0.5% SAP-1 (fine)

Figure 3.16 Materials for in-drilling conditions.

3: Formula 1 + 0.5% carbon fiber (50 mesh) + 1% fiber (100 mesh) + 0.5% fiber (150 mesh)

4: Formula 1 + 2% plugging agent for in-drilling conditions (SAP-1: fiber = 3:1)

5: Formula 1 + 2% plugging agent for in-drilling conditions (SAP-1: fiber = 2:1)

6: Formula 1 + 2% plugging agent for in-drilling conditions (SAP-1: fiber = 1:1)

Table 3.12 Rheological data of different formulas.

No.	Condition	Density (g/cm^3)	AV (mPa · s)	PV (mPa · s)	YP (Pa)	YP/ PV	GEL (Pa/Pa)
1	Before hot rolling	1.4	78	56	22	0.39	2/4
2	Before hot rolling	1.4	85	59	26	0.44	2.5/5
3	Before hot rolling	1.4	84	60	24	0.4	3/5
4	Before hot rolling	1.4	97	70	27	0.39	3/5.5
5	Before hot rolling	1.4	99	74	25	0.34	2.5/5
6	Before hot rolling	1.4	88	66	22	0.33	2.5/5

As can be seen from Table 3.12, after the mixture of CMG and fiber was added to the drilling fluid system, the AV exhibited some increase, but the increase was not large. Its flow state is shown in Fig. 3.17.

The plugging effect of formulas 1, 2, 3, 4, 5, and 6 drilling fluids was evaluated by sand bed test, and the filtration loss was measured every 10 min under 3.5–8 MPa.

It can be seen from Table 3.13 that the addition of fiber and SAP-1 significantly improved the plugging performance of the drilling fluid system. The pressure-bearing capacity reached 8 MPa, and the maximum mesh of sand samples reached 5–10. In addition, the added fiber significantly improved the plugging effect.

3.2.1.2 Evaluation of rheological properties, filtration loss, and plugging effect

The series of SAP plugging materials for in-drilling conditions were added to the drilling fluid system to test their rheological properties, filtration loss and plugging effect at room temperature. The formulas are as follows:

7: 3% bentonite mud + 0.2% NaOH + 0.2% macromolecular emulsion + 3% asphalt + 2% KFT + 0.4% PAC-LV + 2% $CaCO_3$ + 5% KCL + barite

8: Formula 7 + 1% SAP-1 (coarse) + 1% SAP-1 (fine)

9: Formula 7 + 1.5% SAP-2 (coarse) + 0.5% SAP-2 (fine)

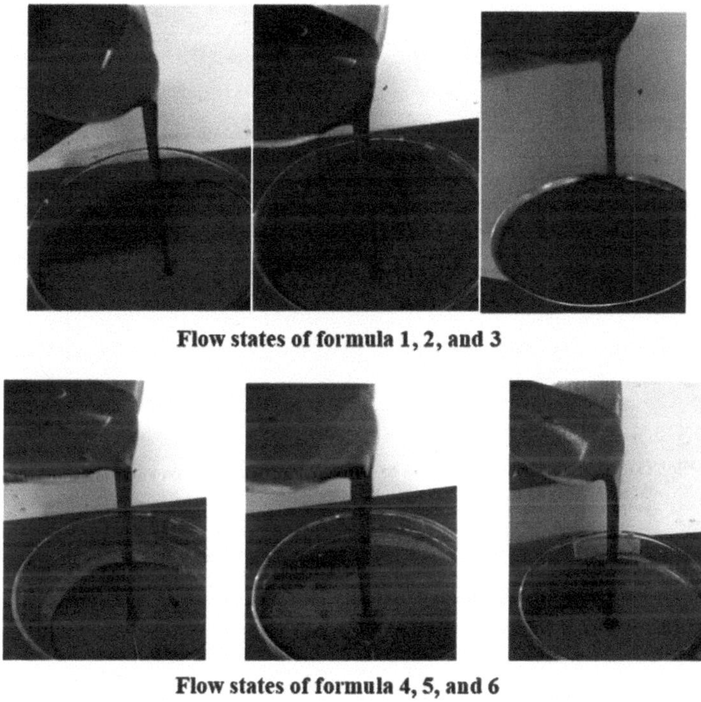

Figure 3.17 The flow state of each formula (formulas 1–6).

3.2.1.2.1 Evaluation of rheological properties and filtration loss

First, 300 mL of drilling fluid (formula 7) was taken and fully stirred. Then a six-speed rotary viscometer was used to test the rheological properties of the drilling fluid. A certain concentration of plugging agent for in-drilling conditions was added according to the volume ratio (formulas 8 and 9). After fully high-speed stirring, a six-speed rotary viscometer was used to test the rheological properties.

It can be seen from Table 3.14 that SAP had little influence on the rheological properties and filtration loss. The results in Fig. 3.18 also showed that SAP had little effect on the flow state of the drilling fluids.

3.2.1.2.2 Evaluation of plugging effect

Through a large number of literature investigations and experimental results, it is shown that the sand bed filtration loss and polymer gel retention capacity with heterogeneous pores can more realistically simulate the downhole situation and reflect the plugging of the underground

Table 3.13 The plugging effect of the drilling fluid with formulas 1, 2, 3, 4, 5, and 6.

Formula	Cuttings sizes (mesh)	Temperature (°C)	3.5 MPa	6 MPa	Maintained at 8 MP for 30 min
1	40–70	90	0	0	0
	20–40	90	Completely lost	–	–
2	40–70	90	0	0	0
	20–40	90	Completely lost	–	–
3	20–40	90	0	0	0
	10–20	90	5	6	6
	5–10	90	48	48	48
4	40–70	90	0	0	0
	20–40	90	Completely lost	–	–
5	40–70	90	0	0	0
	20–40	90	18.5	18.5	18.5
6	40–70	90	0	0	0
	20–40	90	0	0	0

Table 3.14 Rheological properties and filtration loss of the drilling fluids with formulas 7, 8, and 9.

No.	Condition	Density (g/cm^3)	AV (mPa · s)	PV (mPa · s)	YP (Pa)	YP/PV	GEL (Pa/Pa)	FL_{API} (mL)
7	Before hot rolling	1.4	35	32	3	0.09	0.5/1	7
8	Before hot rolling	1.4	41	39	2	0.05	1/1.5	6.6
9	Before hot rolling	1.4	43	31	2	0.06	1/1.5	6.2

formation. The HTHP sand bed plugging experiment is to use the 71-type HTHP water loss meter to fill a certain amount of quartz sand into the container, adjust the particle size of the quartz sand to simulate different formation pores and cracks, and record the sand bed filtration loss to measure the plugging performance of the gel.

The sand bed plugging test evaluation of formulas 7, 8, and 9 drilling fluid was carried out, and the filtration loss at 0.5–5 MPa was measured every 5 min to investigate the plugging effect of the plugging drilling fluid

Flow state of formula 7, 8, and 9

Figure 3.18 The flow state of each formula (formulas 7–9).

Table 3.15 The plugging effect of formulas 7, 8, and 9.

Formula	Mesh	0.5 MPa	1 MPa	1.5 MPa	2.5 MPa	4 MPa	5 MPa
7	40–70	0	0	0	0	0	0
	20–40	Completely lost					
8	20–40	12	17	20	24	35	38
	10–20	Completely lost					
9	20–40	0	6	8	14	23	25
	10–20	35	63	68	78	89	113

system. The data is shown in Table 3.15. SAP plugging agent has a certain plugging effect. For 10–20 and 20–40 mesh sand bed, it can both withstand more than 5 MPa.

The state of the sand bed after plugging is shown in Fig. 3.19. After the plugging agent SAP series is added to the drilling fluid, a solid sand bed can be formed, and the fiber-containing plugging agent SAP has better cementation and plugging effect.

3.2.2 Development and performance evaluation of a high-temperature-resistant supramolecular microgel bionic plugging agent for in-drilling conditions

Under high-temperature conditions, the conventional linear polymer plugging agents easily fail due to chain breakage, which is very common. The cross-linked polymer microgel can form a strong 3D grid structure through chemical cross-linking, leading to good temperature resistance. At the same time, the cross-linked polymer microgel can swell to a certain extent after absorbing water and has elastoplasticity,

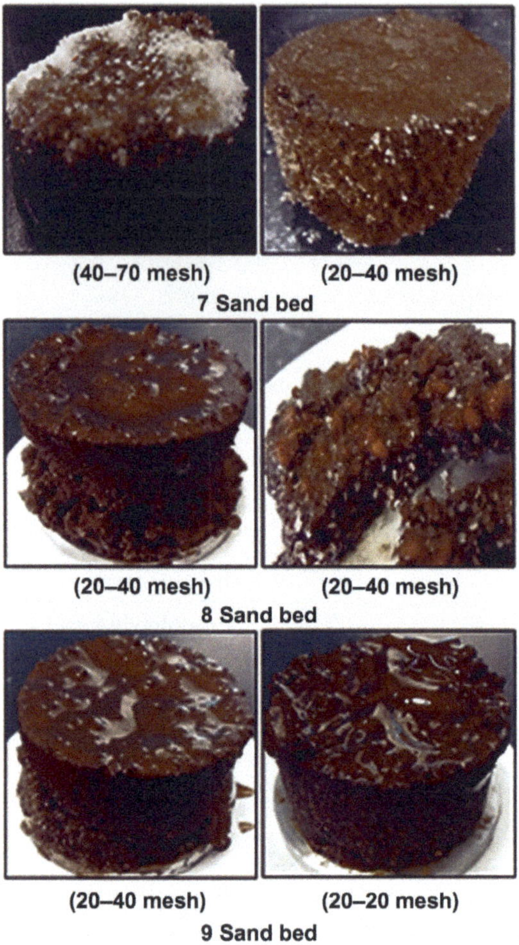

Figure 3.19 The state of the sand bed after plugging.

which enables it to be squeezed into the microcracks of filter cakes to improve their performance, and enter the easy-to-leak formations to achieve plugging under in-drilling conditions. For this, based on the theory of supramolecular chemistry, a bionic nonlinear cross-linked polymer microgel plugging agent for in-drilling conditions has been developed, and the technology has been formed. This technology solves the problem of plugging under in-drilling and high-temperature conditions with little effect on the rheological properties of drilling fluids, that is, the technology has the dual functions of filtration control and plugging under in-drilling conditions.

3.2.2.1 Characterization of the microgel plugging agent

The infrared spectrum and thermogravimetric analysis of the microgel plugging agent (Fig. 3.20) are shown in Figs. 3.21 and 3.22, respectively. In Fig. 3.21, 3408 cm^{-1} corresponds to the stretching vibration peak of the N–H bond, 2926 cm^{-1} corresponds to the stretching vibration peak of the C–H bond, 1660 cm^{-1} is the stretching vibration peak of the C=O bond, and 1452 cm^{-1} is the wagging vibration peak of the C–H bond. There is no vibration peak between 1645 and 1620 cm^{-1}, indicating that there is no C=C bond.

It can be seen from Fig. 3.22 that, the first descending step appears before 250°C, with a mass reduction rate of 4.42%, which is mainly due to the thermal degradation of the stabilizer and other impurities in the microgel. The second descending step is between 250°C and 330°C, which is mainly due to the thermal degradation of the microgel's functional groups, such as amide functional groups. It can be clearly seen that the derivative thermogravimetry (DTG) curve has a peak and valley around 300°C. Since the downhole temperature is usually below 300°C, no further analysis of higher temperatures was conducted.

3.2.2.2 Analysis of morphology and particle size

Transmission electron microscopy (TEM) and scanning electron microscope (SEM) were used to observe the micromorphology of the microgel plugging agent. The results are shown in Figs. 3.23 and 3.24. As can be

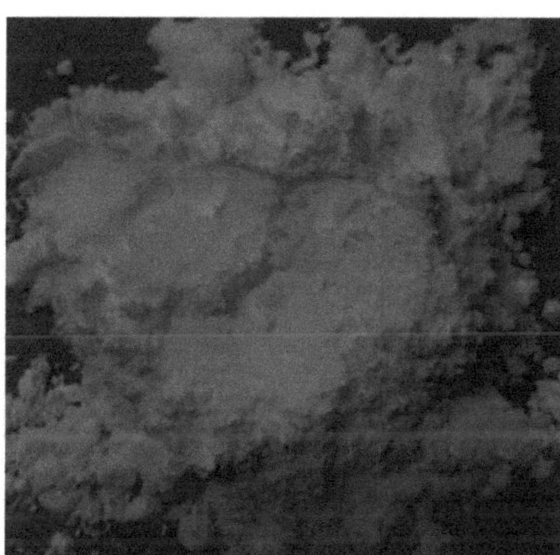

Figure 3.20 Appearance of the microgel plugging agent.

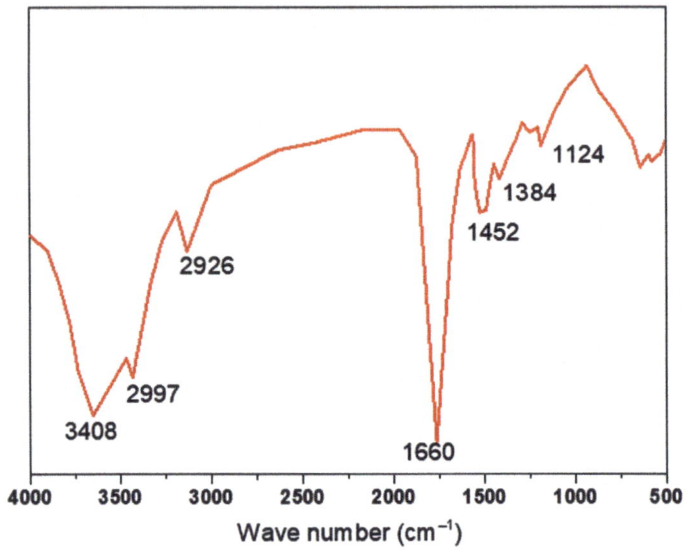

Figure 3.21 Infrared spectrum of cross-linked polymer microgel.

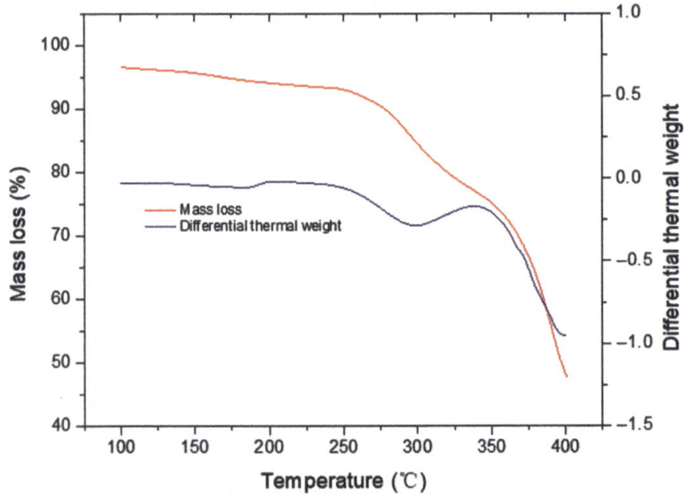

Figure 3.22 Thermogravimetric analysis results.

seen, the microgel mainly consist of spherical or ellipsoidal aggregates, with particle size distributed mainly on a micron scale. The particle size analysis results from Fig. 3.25 show that 95.3% of the particle sizes are distributed in 1–100 μm and 4.7% are in 0–1 μm.

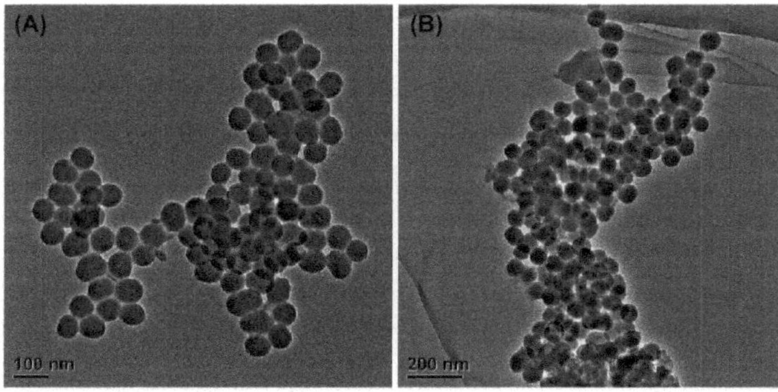

Figure 3.23 TEM analysis of the microgel.

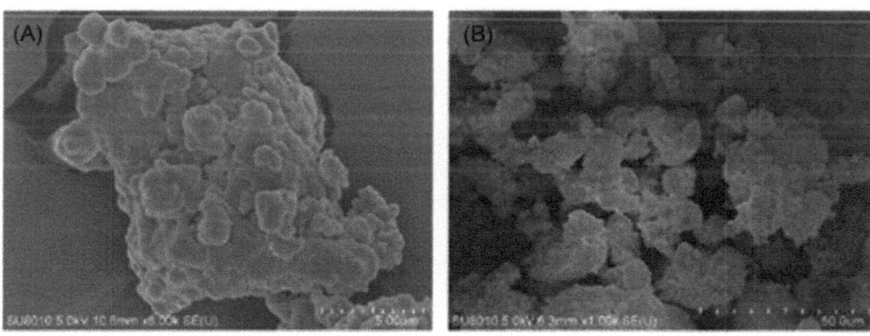

Figure 3.24 SEM analysis of the microgel.

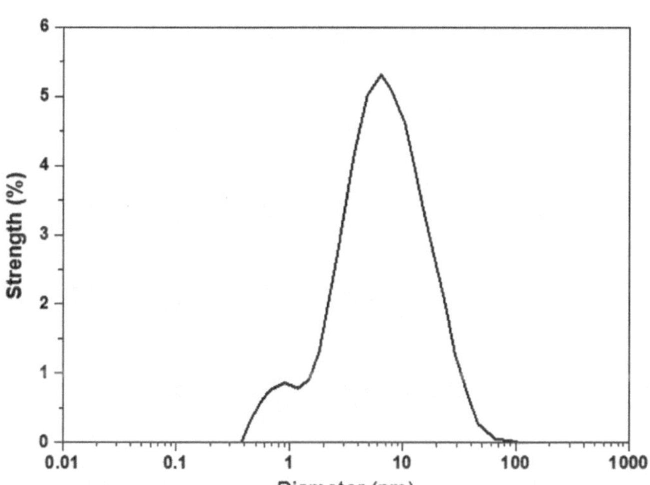

Figure 3.25 Particle size analysis of the microgel.

3.2.2.3 Evaluation of rheological properties

A 4% sodium-based bentonite mud was prepared. After being cured at room temperature for 24 h, it was used as base mud. Different concentrations of microgel plugging agent were added to the base mu, and hot rolled at 150°C, 180°C, and 200°C, respectively. After hot rolling, a six-speed viscometer was used to test the AV, with the results shown in Figs. 3.26 and 3.27. As the microgel concentration increased, the viscosity of the base mud did not increase much. After hot rolling at 180°C and 200°C the viscosity of the base mud even decreased. This is because the microgel plugging agent is

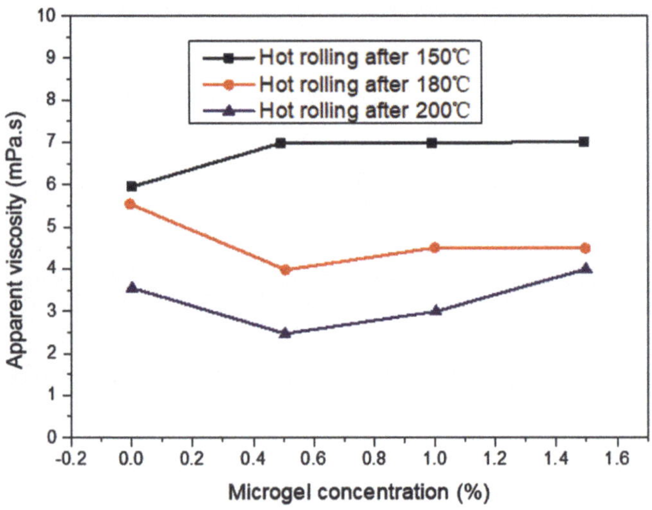

Figure 3.26 Effect of microgel concentration on the AV of base mud.

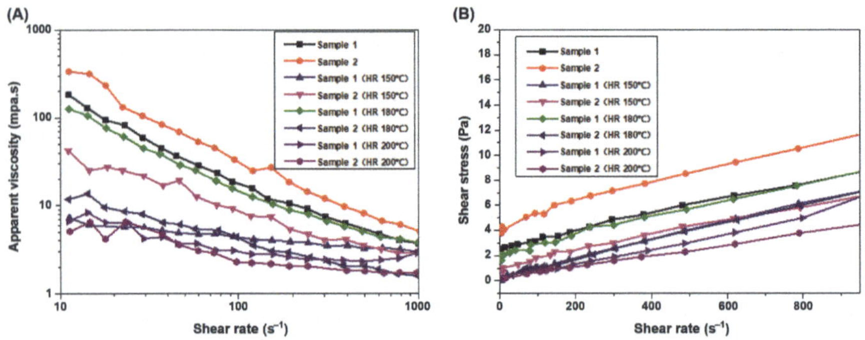

Figure 3.27 Effect of microgel concentration on the rheological properties of base mud.

a nonlinear polymer that did not significantly increase the viscosity of the drilling fluid and could be used as a plugging agent for in-drilling conditions to deal with permeability-type leakage during the drilling process.

3.2.2.4 Evaluation of plugging effect

A 4% sodium-based bentonite mud was prepared. After being cured at room temperature for 24 h, it was used as the base mud. Different concentrations of microgel plugging agent and other treatment agents were added to the base mud to form a high-temperature-resistance supermolecule microgel bionic plugging system for in-drilling conditions, followed by being hot rolled at 150°C, 180°C, and 200°C. After hot rolling, a medium-pressure water loss meter was utilized to test API filtration loss, and the morphology of different sample filter cakes was observed by SEM technique. Fig. 3.28 illustrates the effect of microgel concentration on the filtration loss of base mud. At different temperatures, with the increase of the microgel concentration, the filtration loss of the base mud was significantly reduced. When the microgel concentration increased to over 1.0%, the filtration loss no longer changed obviously, indicating that the microgel plugging agent has good filtration control performance. Fig. 3.29 compares the filtration control effects of different filtration control agents at different temperatures. When the temperature was higher than 120°C, the filtration control effects of PAC

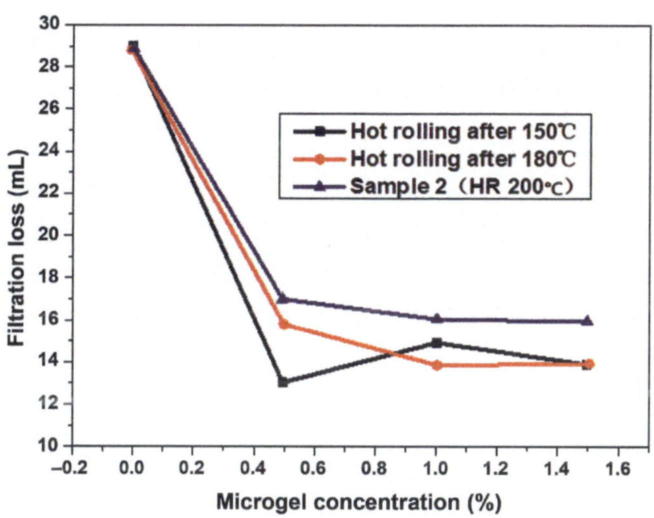

Figure 3.28 Effect of microgel concentration on the filtration loss of base mud.

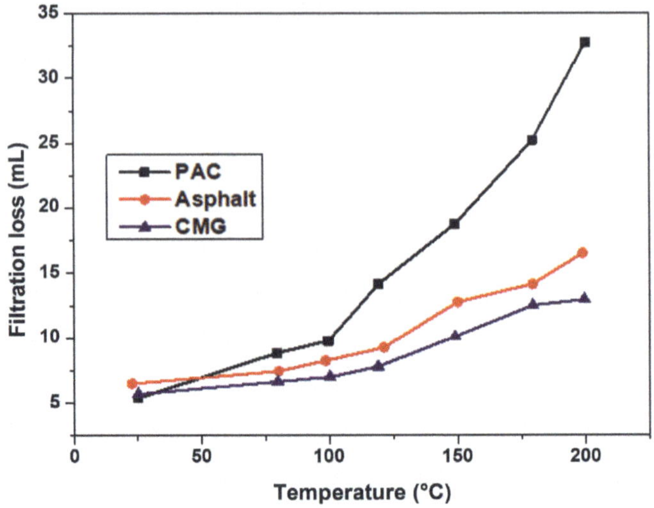

Figure 3.29 Comparison of filtration control effects of different filtration control agents at different temperatures.

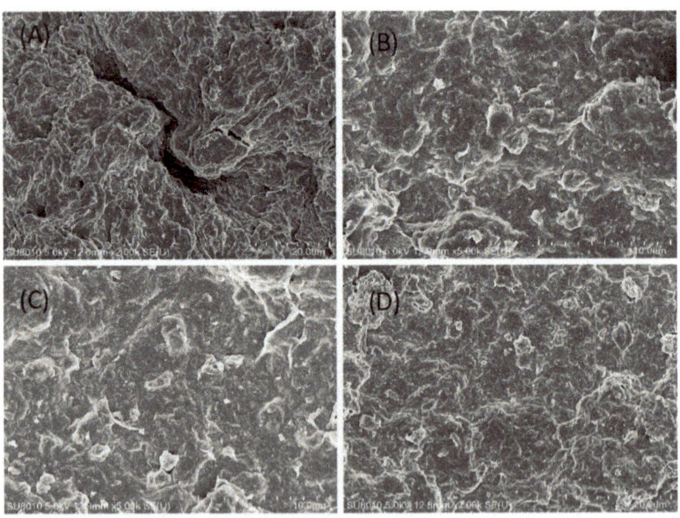

Figure 3.30 SEM images of filter cakes. (A) The filter cake after base mud was hot rolled at 150°C. The filter cake after base mud + 1% microgel plugging agent were hot rolled at (B) 150°C, (C) 180°C, and (D) 200°C.

and asphalt in the base mud were significantly deteriorated, while the microgel plugging agent maintained good filtration control effect, indicating that it had superior temperature resistance ability. From the microscopic morphology of the filter cake (Fig. 3.30) it can be seen that the quality of the filter cake was significantly improved, there were fewer microcracks, and the roughness of the filter cake was reduced after the microgel plugging agent was added. This shows that the microgel plugging agent was effectively filled into the filter cake after absorbing water and swelling.

The high-temperature-resistance supermolecule microgel bionic plugging system for in-drilling conditions is mainly suitable for plugging under in-drilling conditions in microleakage formations but has poor adaptability to medium leakage. For small or medium leakage under in-drilling conditions, the use of strong-adhesion expandable gel-fiber composite plugging technology is recommended. Of course, for formations with large and serious leakage, it is better to conduct plugging under stop conditions.

3.3 Field application of the new technology of bionic drilling fluids for plugging under in-drilling and stop conditions

3.3.1 Introduction

The new technology of bionic drilling fluids for plugging under in-drilling and stop conditions has been widely applied in Slbcopower Company (referred to as SCP Company), the Fourth Branch of CNPC Bohai Drilling Engineering Company (referred to as Bosi Company), Zhundong Drilling Company of CNPC Xibu Drilling Engineering Company (referred to as Zhundong Company), and others. According to incomplete statistics, in 2017 the supramolecular bionic plugging technology was applied to conduct plugging over 20 times in Bohai's Fourth Branch and Xibu's Zhundong Company, with a success rate of 100%. At the same time, the technology was introduced by SCP Company, which solved the traditional series of lost circulation problems. Data for some construction wells using this technology are shown in Table 3.16. Among the three wells of Xizuanke 205, Xizuanmahu 016, and Xizuan Ma21021 the failure of the traditional plugging method has resulted in the loss of about 1000–2400 m^3 of drilling fluid. The direct economic loss is more than 3 million CNY (approximately 1100 CNY/m^3 of drilling fluid). Later, this technology was used for plugging. The construction cost was about 210,000 CNY. The plugging was very successful, and the effect was remarkable.

Table 3.16 Use of some bionic plugging technologies.

Company and drilling crew	Amount of lost circulation (m^3)	Amount of bionic drilling mud (m^3)	Plugging effect
SCP-YB025-D01	378	30	Success
SCP-YB361-H03	447	30	Success
SCP-YB025-D01	775	30	Success
Bosiliu 107-72x	216	25	Success
Bosiliu 107-76x	287	25	Success
Bosidaji6-2xiang5	765	30	Success
Su 25-18-29X	417	25	Success
Xizuanke 205	2452.74	40	Success
Xizuanmahu 016	1368	50	Success
Xizuan Ma21021	1076	40	Success

Note: Amount of lost circulation refers to the total amount of lost circulation caused by the failure of other plugging technologies. The bionic plugging technology was adopted after the failure of other plugging technologies.

3.3.2 Typical cases

3.3.2.1 Successful plugging of Well Ke205

Well Ke205 is located in Well 80-Mahu 1 on the west slope of Mahu, the northeast wing of Zhongguai bulge in the western uplift of the Junggar Basin. The drilling depth was 4496 m, and drilling was designed to be completed in three spuds. The three-spud drilling used a potassium-calcium-based polymer drilling fluid system. When drilled to the Baijiantan formation, returns and serious lost circulation occurred at multiple places (at 2939 m). The Baijiantan formation is about 312 m thick. The upper part is gray unequal granite; the lower part is dark gray shale and argillaceous siltstone; and there are longitudinal cracks. When lost circulation occurred, the drilling fluid density was 1.26−1.32 g/cm^3, the viscosity was 47−60 s, and the drilling fluid displacement was 20 L/s. The CNPC Xibu Drilling Engineering Company invited a professional plugging company to implement crack-plugging construction using other plugging technologies. The plugging slurry with a concentration of 10% was prepared, and 40 m^3 was injected into the bottom of the well, but the drilling fluid displacement was 20 L/s, and the casing pressure was always 0 MP. A total of 2452.74 m^3 of drilling fluid was lost, which was a failure.

Owing to the high-porosity and high-permeability formation in Well Ke205 and the low formation pressure coefficient and fracture pressure, induced lost circulation easily occurs. The advantages of supramolecular gel and gellant were utilized, such as high strength, fine particles, and irregular shapes. The materials were wedged into the leakage channel. Because it was not easy for them to flow away, high-strength plugged zones were formed. Supramolecular bionic crack plugging was implemented, and the leakage zone was plugged. A total of 40 m^3 of supramolecular bionic plugging slurry (density of 1.30 g/cm^3, viscosity of 60 s) was prepared, and 20 m^3 was wedged into the well, 3.1 m^3 was squeezed, the drilling fluid displacement was 45 L/s, and no lost circulation occurred. So the plugging was successful, and drilling was resumed. The formula of plugging slurry was base mud + 1.5% supramolecular gel + 5% type II gellant + 7% type III gellant + 6% type IV gellant + 7% type V gellant.

3.3.2.2 Successful plugging of Well Zhundong MA21021

Well Ma21021 is located in the Mabei slope belt of the west ring belt of the Mahu sag in the Central Depression of the Junggar Basin. The well depth is 3645 m, and the drilling was designed to be completed in two spuds, with a potassium-calcium-based polymer system for drilling. When drilling reached the Badaowan Formation (beyond 2459 m), serious lost circulation occurred. There were longitudinal cracks in the Badaowan Formation with a total thickness of about 665 m. The upper part is gray unequal granite, and the lower part is dark gray shale and argillaceous siltstone. The leakage zone is a high-porosity and high-permeability formation, with low formation bearing pressure, low formation fracture pressure, and low pressure coefficient. The formation contains gravel, which can easily cause induced lost circulation. When lost circulation occurred, the drilling fluid density was 1.23−1.26 g/cm^3, and the viscosity was about 53−58 s. GRANDOIL's specialized plugging company used other plugging technologies to implement plugging construction. After 30 m^3 of plugging slurry was injected, the plugging failed, and a total of 1076 m^3 of drilling fluid was lost. After the plugging failed, the supermolecular bionic gel type I plugging slurry was used for plugging construction, and 28 m^3 was injected. The advantages of supramolecular gel and gellant were utilized, such as high strength, fine particles, and irregular shapes. The materials were wedged into the leakage channel. Because it was not easy for them to flow away, high-strength plugged zones were formed, achieving the purpose of plugging the leakage zone, and the plugging was successful. The formula of plugging slurry was base mud + 1.5% supramolecular gel + 6% type I gellant + 25% walnut shell.

3.3.2.3 Successful plugging of Well Daji 6-2 Xiang 5

Well Daji 6-2 Xiang 5 is located in the northern part of Nanshan Village, Xujiadu Town, Daning County, Linfen City, Shanxi Province, China, and is in the east of the Yishan Slope in the Ordos Basin, with a well depth of 2534.89 m. The drilling was designed to be completed in two spuds using polysulfonate drilling fluid. The drilling fluid had a density of 1.18 g/cm^3 and a viscosity of 55 s. When drilled to the depth of 1174 m, it reached the Shiqianfeng formation and serious lost circulation occurred, but the position of lost circulation was uncertain. The first section of the Shiqianfeng formation is about 200 m thick, consisting of brown-red shale and light red, brown-gray, and light-gray slender stone sandstone, with a low pressure coefficient. So it is a high-porosity and high-permeability formation and is likely to cause induced lost circulation. When the lost circulation occurred, the density of the drilling fluid was 1.13 g/cm^3, and the viscosity was 40 s. At first, other technologies were used to implement the plugging but failed. It was unable to continue drilling, and a total of 765 m^3 of drilling fluid was lost. After the plugging failed, bionic plugging technology was used for plugging. In this case, 30 m^3 of type III plugging slurry was prepared, and 19 m^3 of supramolecular plugging slurry was squeezed. The slurry return was normal, the plugging was successful, and drilling was resumed.

The formula of the plugging slurry was 10 m^3 well mud + 5 m^3 clear water + 15% type III gellant + 1.5% type I gel and 10 m^3 base mud + 5 m^3 clear water + 5% type IV gellant + 1.5% type I gel.

3.3.2.4 Successful plugging of Well Liu107-72X

Well Liu 107-72X is located 200 m south of Sishang Village, Shisu Town, Suning County, Ning 68 Fault Block, Suning Oilfield, Raoyang Sag, with a well depth of 3673 m. The drilling was designed to be completed in two spuds. The polysulfonate drilling fluid was used, with a density of 1.25 g/cm^3 and a viscosity of 52 s. After drilling to the depth of 3406–3473 m, multiple places of lost circulation were found, at which time the drilling fluid density was 1.21 g/cm^3, the viscosity was 49 s, the displacement was 30 L/s, and the formation was lithological gray shale. The leakage zone was in the first section. The lithological gray shale is mixed with siltstone and dolomite. The cause of the lost circulation may be that the sand shale was not strongly cemented, resulting in differential pressure lost circulation or formation deficit. To avoid further lost circulation, a total of 94 m^3 of plugging slurry was prepared to implement plugging, but the plugging failed, resulting in the loss of 216 m^3 of drilling fluid throughout the well. After the failure, supramolecular bionic gel was used for plugging, which was successful. The formula of the leakage slurry was 15 m^3 base mud + 20% type IV gellant + 1.5% type I supramolecular gel.

3.3.2.5 Successful plugging of Well Liu107-76X

Well Liu 107-76X is located 450 m south of Sishang Village, Shisu Town, Suning County, Hebei Province, and it belongs to the Ning 68 Fault Block of Suning Oilfield. The drilling was designed to be completed in two spuds, with a depth of 3716 m. The polysulfonate drilling fluid system was used, with a density of 1.24 g/cm^3 and a viscosity of 40 s. During the drilling of the adjacent Well Ning 68 in Well 107-76X, a total of 22 places of lost circulation occurred in the section from 3085.39 to 3819 m. A total of 2743 m^3 of drilling fluid was lost, and there were cracks. It was judged that there might be serious lost circulation in Well 107-76X. After completion of drilling, the 40-column pumping cycle was started, and the circulation displacement was 23 L/h. After 40 min of circulation the displacement was increased to 32 L/h, at which time return-type lost circulation occurred suddenly, with a drilling fluid density of 1.24 g/cm^3 and a viscosity of 45 s. The leakage zone is located at Dongying formation–Shahejie formation. Dongying formation is about 990 m thick, and the upper part is purple-red shale and brown-red and light gray fine sandstone with unequal thickness interbedded. After the lost circulation occurred, traditional plugging was carried out three times, but all attempts failed. A total of 287 m^3 of drilling fluid was lost. Then bionic plugging technology was used. There was no lost circulation during the process of replacing 13 m^3 in the well with 20 m^3. After 6 m^3 of supermolecular bionic plugging slurry was squeezed in, there was no lost circulation, and the drilling continued smoothly. So the plugging was successful. The formula of leakage slurry was base mud + 1.5% supramolecular gel + 5% type V gellant + 10% type IV gellant + 5% type III gellant.

References

[1] F. Shengli, L. Jun, Complex downhole conditions and the corresponding countermeasures, Drill. Prod. Technol. (2004).

[2] X. Yongjie, Study on the Formation Stress Calculation Method and Well Leak Mechanism for Aloft-steep and Complex Structure, Southwest Petroleum University, 2005.

[3] S. Fei, Z. Dalin, W. Mingquan, Research on the countermeasures of lost circulation analysis and treatment technology, Neijiang Sci. Technol. (3)(2008) 109. 147.

[4] Y. Xinbin, Research on Leakage Prevention and Plugging Technology in Yingtai District of Daqing, Northeast Petroleum University, 2003.

[5] W. Linzhong, The Study on the Prevention and Plugging Flow Technology in Beisi District, Northeast Petroleum University, 2007.

[6] N. Sizhao, Study on the Newly Drilling Fluid Technology for Lost Circulation Prevention and Control, China University of Petroleum, 2007.

[7] L. Jiaxue, H. Jinjun, L. Pingya, W. Shuqi, H. Tao, L. Xingshu, Study on the technology of leak prevention and plugging for in-drilling condition, Drill. Fluid Completion Fluid (2008).

[8] X. Jiyou, C. Zhong, X. Liang, et al., Research and application of leak resistance and sealing technology while drilling, Drill. Prod. Technol. (2)(2007) 7−10. 19, 151.

[9] H. Tao, Analysis and research on plugging technologies and materials during drilling construction, Sci. Technol. Inf. (2013).

[10] Z. Hongli, G. Yan, W. Zhilong, Lost circulation materials in China, Spec. Oil Gas. Reserv. (2)(2004) 1−2. 10−96.

[11] W. Jun, X. Mingbiao, C. Lin, L. Chao, Laboratory research of a new type of oil-based plugging agent, J. Yangtze Univ. (Nat. Sci. Ed.) 13 (11) (2016) 46−48. 4−5.

[12] R. Mao, S. Junlin, F. Zhou, et al., Analysis of antileak and sealing technology, Drill. Prod. Technol. 32 (03) (2009) 29−30. 125.

[13] J. Bin, W. Tao, Research on the process technology of oil and gas well leakage in the western Sichuan area, China Pet. Chem. Stand. Qual. (2013) 1−15.

[14] L. Jun, Modified Study on Experimental Technique for Lost Circulation Evaluation, China University of Petroleum, 2010.

[15] W. Deyu, P. Xiaolin, S. Taihe, L. Xitan, Z. Jianqing, DL-1 type plugging test device and evaluation method, Oil Drill. Prod. Technol. (5)(1996) 44−48. 107.

[16] L. Kaihe, Study and Application of Lost Circulation Resistance and Control Technology During Drilling, China University of Petroleum (East China), 2007.

[17] Z. Xionghu, C. Shengyuan, Feasibility study of sand bed method to evaluate filtration loss of drilling fluid, West-China Drill. Eng. 5 (2009) 83−84.

[18] W. Qing, Research on Leak Prevention and Plugging Technology in Zhangwu and Jiulongshan Areas, Southwest Petroleum University, 2015.

[19] Z. Yipeng, Processing method of downhole complicated problem in the process of air drilling, Chem. Intermed. (5)(2016) 90−91.

[20] Z. Wei, Study on Environmentally-Friendly Drilling Fluid for Leak Resistance and Sealing, China University of Petroleum, 2008.

[21] W. Qiucheng, Z. Longhua, Analysis of the cause of lost circulation and discussion on geological application, Mud Logging Eng. (3)(1996) 43−44.

[22] S. Lijia, Research on the Lost Circulation Control in Lincan-1 Well of Jilin Oilfield, Northeast Petroleum University, 2015.

[23] H. Sanqing, Model JLX-2 dynamic loss plugging tester, China Pet. Machinery 28 (6) (2000) 13−14.

[24] Z. Xiwen, L. Shuang, Z. Jie, S. Jinsheng, Y. Zhi, Research progress on lost circulation materials and lost circulation control technology, Drill. fluids completion fluids 26 (6) (2009) 74−76. 79, 97.

[25] J. Shaoqing, Z. Zhaoxia, Z. Huisheng, J. Changsheng, H. Junyu, Q. Jiazheng, Laboratory study of a new ultra-low density lost circulation control fluid, Drill. fluid completion fluid 26 (6) (2009) 31−32. 36, 94.

[26] X. Xiaoyue, Lost circulation analysis and treatment technology, China Petrochem. (3)(2017) 75−76.

[27] W. Linzhong, The Study on the Prevention and Plugging Flow Technology in Beisi District, Daqing Petroleum University, 2007.
[28] W. Zhonghua, The status and development direction of plugging technology for complex formation lost circulation, Sino-Global Energy 19 (1) (2014) 39–48.
[29] H. Songquan, Discussion on the types of borehole lost circulation and plugging technology, Geol. China (5)(1991) 17–19.
[30] G. Baoli, Y. Menglei, A Simulation Test Device for Drilling Plugging: China, 02243296.5, 2003, pp. 9–24.
[31] D. Mingxiang, BDY-1 portable plugging instrument for drilling fluid, Drill. Fluid Completion Fluid 19 (2) (2002) 30–31.
[32] L. Yingsong, J. Jinbao, Q. Tao, The development of well loss processing technology, Fault Block Oil Gas. Field (2)(2005) 4–7. 89.
[33] L. Yanqiang, X. Tongtai, Y. Zhenjie, G. Zhimin, Recent progress on preventing and treating lost circulation domestic and overseas, Drill. Completion Fluids 27 (6) (2010) 80–84. 102.
[34] T. Yi, W. Zongyan, W. Ke, F. Qi, Structural bionic design for tower cranes boom based on king lotus leaf vein branched structure, Mech. Des. Manuf. (3)(2017) 36–39.
[35] L. Yan, Z. Jie, W. Lai, L. Fengmin, Mechanical properties and toughening mechanisms of mollusk shell, J. Mech. Strength. (3)(2007) 507–511.
[36] Z. Xin, Occurrence and Bleaching Experiment of Organic Matter in Gem-Grade Red Coral, China University of Geosciences, 2012.
[37] L. Di, Chemistry and gems, N. Curric. Tutor. (11)(2012) 11.
[38] T. Jun, Preparation and Properties of B_4C/BN Layered Composite Ceramic Materials, Jingdezhen Ceramic Institute, 2008.
[39] G. Zhongliang, X. Xiandue, T. Jianping, Y. Zhong, L. Dongping, Experimental study of the load capacity of human blood, Mech. Sci. Technol. (11)(2006) 1261–1263.

CHAPTER 4

New technology of bionic drilling fluids with ultralow formation damage

Normally, varying degrees of formation damage occur during the exploration and development of oil and gas resources, resulting in decreased production or even scrapped formations. Drilling fluid, which is the first foreign fluid to come into contact with the formation, may cause extensive damage to the formation. To reduce or avoid the damage caused by drilling fluids and to increase the production of single wells, global scholars have conducted extensive research for more than half a century and have established three generations of temporary plugging technology for drilling fluids: shielding temporary plugging, fine temporary plugging, and temporary plugging with physicochemical film. of formation protection effects have been gradually improved, along with obvious economic benefits [1–32].

The technical idea of shielding temporary plugging is to add the bridging particles (content greater than 3%) with 1/2–2/3 pore throat diameter to the drilling fluid, the filling particles (content greater than 1.5%) with about 1/4 pore throat diameter, and the deformable particles whose softening point is compatible with the formation temperature and whose particle size is equivalent to the filling particles (the content is generally 1%–2%) to the drilling fluid. Then, in a very short time, the positive pressure differential is used to make the bridging particles, filling particles, and deformable particles enter the pore throat of the formation, forming a plugged zone with a permeability close to zero near the borehole. This can prevent subsequent solid and liquid phases from continuing to invade the formation and cause damage. After drilling is completed, the plugged zone is released by methods such as perforation, chemical dissolution, and flowback to restore oil and gas flow. The essence of this technology is to transform unfavorable factors such as pressure differential and solid phase into favorable factors that benefit the protection of formations, which is actually a transformation of contradictions [33–49].

Fundamentals and Applications of Bionic Drilling Fluids
DOI: https://doi.org/10.1016/B978-0-323-90293-9.00004-6

Owing to the strong heterogeneity of formations, the wide distribution range of pore throat diameters, the small number of large pore throats, and the large number of small pore throats, the theoretically calculated average pore throat diameter will be much smaller than the actual one in the formation if all these pore throats are considered in the calculation. If selecting the particle size of the temporary plugging agent according to such calculation results, the large pore throats will inevitably not be effectively plugged, and many small pore throats will be plugged, leading to poor plugging effect [50–64]. To solve this problem, the team led by Professor Guancheng Jiang of China University of Petroleum (Beijing) innovatively proposed a new idea of simultaneously adding several bridging particles of different particle sizes and types to the drilling fluid in 1999. Guided by this idea, researchers further developed different types of new composite materials for temporary plugging and then developed a series of new technologies of drilling fluids for protecting formations, that is, fractal geometry temporary plugging technology, D90 ideal filling technology, general-purpose temporary plugging technology, alkaline microfiber temporary plugging technology, D50 temporary plugging technology, and so on. These technologies can form a shielding temporary plugged zone with a permeability almost equal to zero near the wellbore with a wider pore throat diameter distribution, which expands the plugging range, and improves the effect of formation protection. Further, a series of theories were developed, such as ideal filling theory, D90 theory, fractal geometry temporary plugging theory, and general-purpose temporary plugging theory, which transformed the technology of drilling fluid with formation protection from the first generation of coarse shielding temporary plugging technology to the second generation of fine shielding temporary plugging technology. Under the same conditions, the protection effect of the second-generation technology is better, and it can help to increase the daily production of a single well.

Whether shielding temporary plugging or fine temporary plugging is used, the particle size of bridging particles needs to be selected according to the pore throat diameter distribution of the formation, so the pore size distribution of the formation needs to be accurately predicted. However, the premise of accurately predicting the pore size distribution of the formation is to analyze and test the cores of all well sections of the formation. In fact, it is impossible to perform core testing without completing the drilling and coring the well. In the actual operation process, the method of using the pore size distribution of adjacent wells or adjacent blocks is used for predicting that of a given well, which is not accurate. Therefore the selection of particle size for temporary plugging agent is bound to be inaccurate, resulting in a good temporary plugging effect for a certain well section but a poor effect for the entire formation. This affects the protection of the formation and is a serious shortcoming of the first and second generations of drilling fluid technology with formation protection. The only way to solve this technical problem is to establish a set of technologies that can effectively prevent the invasion of drilling fluid without accurately predicting the pore size distribution of the formation and thus protect the heterogeneous formations. To this end, Jiang [29] and Sun et al. [65] combined the physical and chemical plugging methods to establish the oil film method and film-forming method, respectively. By using the physical and chemical interaction of oil film agent or film-forming agent molecules, a film is formed on the outer layer of the borehole to shield the borehole and drilling fluid, thereby preventing the filtrate and solid phase from entering and preventing the formation. Then the oil film and film-forming temporary plugging technologies of drilling fluids with formation protection were formed. These technologies achieved good formation protection effects and high production in thousands of wells in oilfields around the world and solved the technical problem of the need to accurately predict the size and distribution of the pore throat of formations for temporary shielding and fine temporary plugging, which has become a relatively active research direction. These two technologies have become representative of the temporary plugging technologies with physicochemical film, which has promoted the transition of the drilling fluid technology with formation protection from physical dense compaction of inert particles to the third generation of temporary plugging technology with physicochemical film. This is great technological progress in the field of drilling fluids with formation protection.

Although the oil film and film-forming technologies have achieved a good transition from physical temporary plugging to temporary plugging with physicochemical film and have solved the serious shortcomings of the need to accurately predict the pore throat diameter distribution of formations, the theoretical basis of establishing temporary plugging technology with physicochemical film is the theory of water-based drilling fluid's film formation on the surface of shale rock. Global researchers have not theoretically confirmed that water-based drilling fluids can indeed form highly efficient films on the rock surface of formations (such as sandstone). Van et al. [55] confirmed that the efficiency of films formed by water-based drilling fluids on shale is much less than that of films formed by oil-based drilling fluids [66–79], and its shielding effect is therefore far inferior to that of the latter. Under normal circumstances the strength of a film is much lower than the strength of the shield ring formed by the accumulation of particles. Large pressure differentials (e.g., pressure surge, multiple pressure systems) easily damage the film and fail to protect the formation, especially high- and ultrahigh-permeability formations. It takes some time to form a film through physical and chemical action. Before a film is formed, owing to instantaneous filtration and other reasons, even a small amount of foreign fluids will still cause damage to the formation, especially the low- and extra-low-permeability formations dominated by the damage mechanisms of water sensitivity and capillary. Neither oil film nor film formation technology can solve the problem of formation damage caused by instantaneous filtration [80–83]. Therefore although the temporary plugging technology utilizing physicochemical film has made great progress in recent years, there is still a large gap with the target of ultralow or zero damage. To this end, Guancheng Jiang et al. introduced bionics for the first time into the field of drilling fluids for protecting formations. By imitating the structure and composition of superamphiphobic nanocrystals on the peristome surface of nepenthes, a superamphiphobic modifier FCS-08 was developed, and a new technology of superamphiphobic bionic drilling fluids for protecting low-permeability formations was established. Imitating the surface structure and characteristics of lotus leaves, a film-type amphiphilic polymer formation protector (i.e., LCM-8) was developed, and a new technology of bionic drilling fluids for protecting medium-permeability formations was established. Taking shell multilayer composite structure featured by alternating mud and brick as a model, a new material with structural bionics was invented using rigid particles and organic amine fiber amphiphilic polymer filming agent, and a new technology of drilling fluids with synergistic effect for protecting high-permeability formations was established. Therefore the theory and technology of the fourth generation of bionic temporary plugging technology of drilling fluids with formation protection abilities have been developed and have achieved satisfactory effects in field applications.

4.1 New technology of bionic drilling fluids to prevent low-permeability formation damage by imitating the superamphiphobic property of the peristome surface of nepenthes

Normally, the liquid phase damage caused by capillary forces is the main form of damage to low-permeability formations. If the wettability of the rock surface of the borehole can be reversed to be hydrophobic and oleophobic (i.e., gas wettability), the contact angle θ of water and oil on the rock surface is greater than 90 degrees and $\cos \theta < 0$, so the capillary suction is changed to capillary resistance. As a result, the liquid phase cannot enter the pore throat of the formation, which avoids liquid phase adhesion and water blocking damage, avoids the occurrence of the main damage factors of low-permeability and ultralow-permeability formations, and helps to increase the production of oil and gas wells. Under the guidance of this theory, bionics was introduced into the field of drilling fluids. By using the superamphiphobic peristome surface of nepenthes as a model, FCS-08 was developed, in which the agent was made from fluorine-containing acrylate polymers. Then the new technology of superamphiphobic bionic drilling fluids for protecting low-permeability formations was formed and has been successfully applied in petroleum engineering [84—87].

4.1.1 Preparation of FCS-08

The fluorinated monomer and emulsifier were slowly dropped into 170 mL of deionized water to produce a uniform emulsion using a shear emulsification method. The product was then transferred to a four-necked flask equipped with an electric stirrer, condenser, and nitrogen introduction device and was subjected to a water bath at 80°C. The initiator was added dropwise to the system in two portions, and the reaction was ended after 3 h. The obtained emulsion was naturally cooled to below 60°C, the pH was adjusted to neutral, and the emulsion was cooled to below 50°C to stop stirring. The emulsion was filtered out to obtain FCS-08 (Fig. 4.1).

4.1.2 Characterization and performance evaluation of FCS-08

4.1.2.1 TEM observation

FCS-08 is composed of two kinds of nanomaterials with different structures, as shown in Fig. 4.2. The synthesized FCS-08 circumscribes many small spheres with a diameter of less than 100 nm in the long structure, and its length and width form a nano-micro structure. The further

Figure 4.1 Sample of FCS-08.

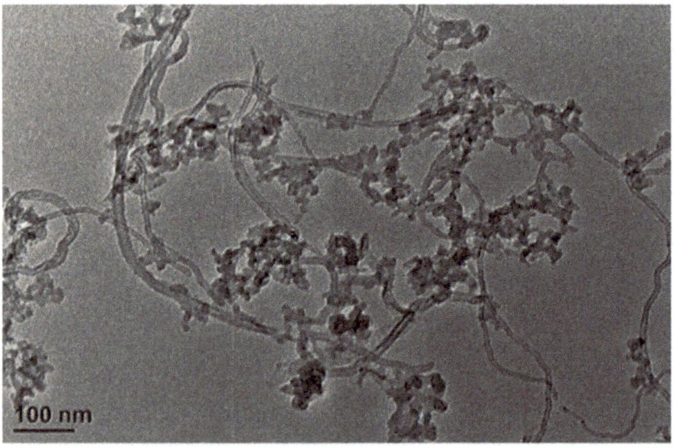

Figure 4.2 TEM image of FCS-08.

aggregation of nanomaterials forms a more complex multilevel structure. This multilevel structure produces more small grooves in the middle, thereby capturing a large number of air bags to prevent the surface from contacting the liquid phase to achieve the amphiphobic effect.

4.1.2.2 SEM observation

The microscopic morphology of the core surface before and after treatment with FCS-08 was observed by the scanning electron microscope (SEM) technique. The results are shown in Fig. 4.3. The surface of the

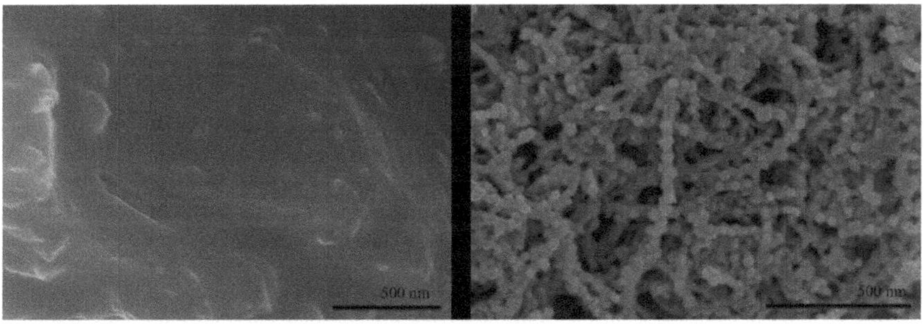

Figure 4.3 SEM images of the surface of the core before (*left*) and after (*right*) treatment with FCS-08.

untreated core is very smooth on the nano-micro scale, and there are only a few convex structures with a small surface roughness. Therefore the original core surface exhibits strong hydrophilic and lipophilic wetting characteristics. When FCS-08 was used to treat the core surface, a series of bead-like structures were stacked on the surface of the core layer. The existence of this special structure and the special chemical properties of FCS-08 itself made the surface exhibit amphiphobic characteristics.

4.1.2.3 Influence of FCS-08 on surface wetting performance

4.1.2.3.1 FCS-08 improves the contact angle of the core surface

It can be seen from Fig. 4.4 that when the core surface was not treated, the water contact angle (polar solvent) on the core surface was 34 degrees, and that of n-hexadecane (nonpolar solvent) was 0 degree. When the mass concentration of FCS-08 was 0.1%, the water contact angle was 118.55 degrees, and that of n-hexadecane was 98.58 degrees. When the concentration of FCS-08 increased to 0.3%, the water contact angle was increased to 160.23 degrees, and that of n-hexadecane was increased to 133.47 degrees. So the surface of the core realized hydrophobicity and oleophobicity. With the concentration of the FCS-08 continuing to increase, the contact angle of the two phases basically remained constant. At the same time, as can be seen from Fig. 4.5, both the water and n-hexadecane droplets gradually changed from a small flat sphere to a large hemisphere after treatment with FCS-08. At the same time, the decreasing contact area between the liquid and the core surface indicated that the wetting performance of the core surface had changed from lyophilic to lyophobic.

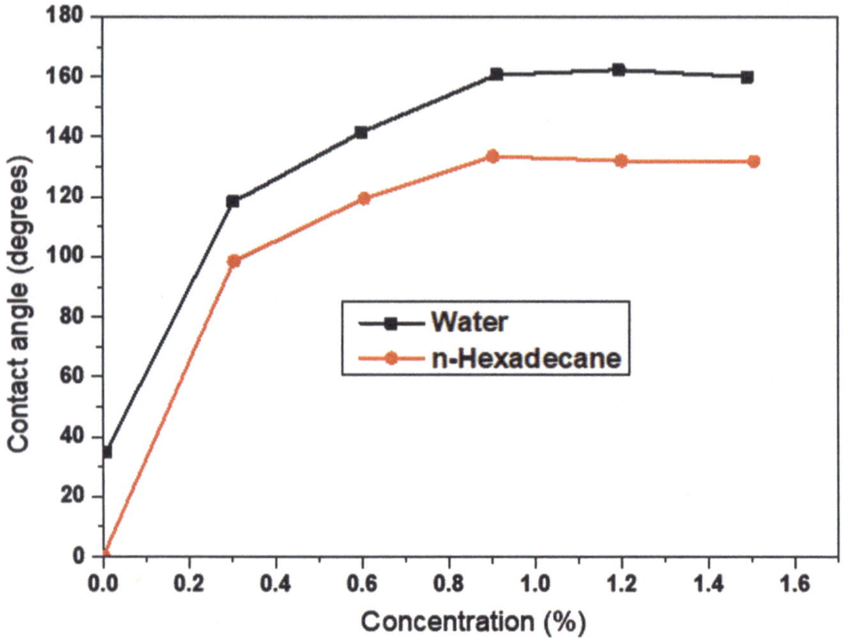

Figure 4.4 Effect of FCS-08 on the two-phase contact angle of the core surface.

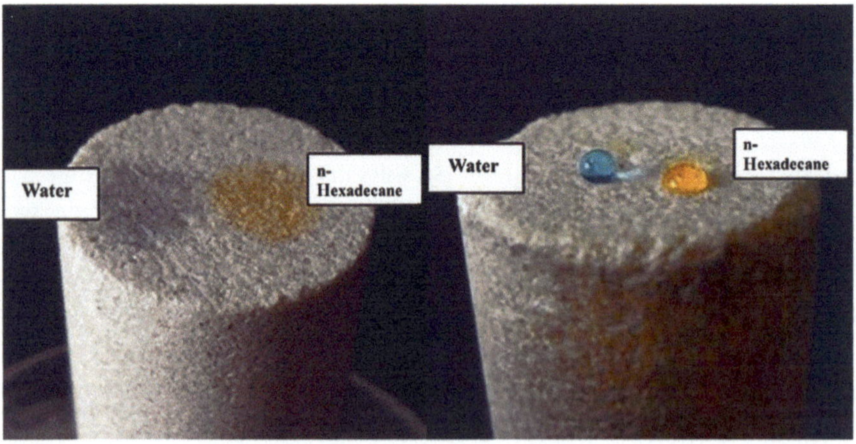

Figure 4.5 Wettability of the core surface before and after treatment with FCS-08.

4.1.2.3.2 The effect of FCS-08 on the surface energy of the core

Based on the measured water and oil contact angles, the surface free energy of the artificial core after the treatment with FCS-08 was calculated by using the Owens two-liquid method. The results are shown in

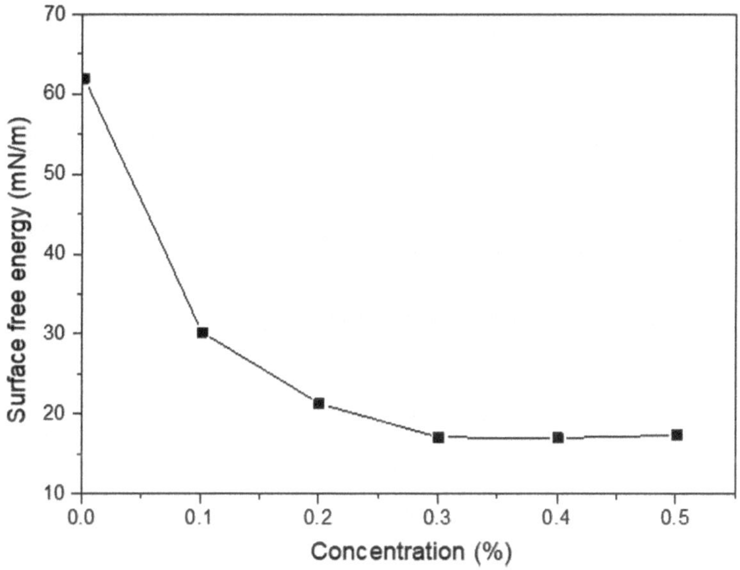

Figure 4.6 Impact of FCS-08 on the surface energy of core.

Fig. 4.6. The surface free energy was determined by the roughness of the material surface and the surface chemical properties. After surface modification by FCS-08, the surface free energy of the core dropped sharply. As the concentration of FCS-08 increased, the core surface free energy continued to decrease. When the concentration increased from 0% to 0.1%, the core surface free energy decreased from 61.80 to 6.57 mN/m. When the concentration continued to increase to more than 0.3%, the core surface free energy dropped below 0.80 mN/m, so the core surface changed from a lyophilic state to a low-surface-energy lyophobic state.

4.1.2.3.3 Surface tension and interfacial tension

The surfactants such as ABS, ABSN, OP-10, TWEEN-80, and SPAN-80, which are commonly used antiwater blocking agents in drilling fluids, and SYP-1, which is a poly alcohol antiwater blocking agent commonly used in acidizing or fracturing fluids, were compared with FCS-08 aqueous solution in terms of the surface tension. The surface tension was measured by using the hanging drop method with a contact angle measuring instrument JC2000D3. The experimental temperature was 25°C. The experimental results are shown in Table 4.1. The interfacial tension of 0.1%–0.5% ABSN and FCS-08 aqueous solutions with oil was compared. The instrument used was the XZD-SP rotating drop ultralow interfacial

Table 4.1 Change of surface tension.

Sample	Surface tension (25°) (mN/m)
0.4% ABS	33.85
0.4% ABSN	33.25
0.4% OP-10	30.65
0.4% TWEEN80	35.15
0.4% SPAN80	25.67
0.4% Polymeric alcoholSYP-1	56.93
0.4% FCS-08	18.40

Table 4.2 Changes in physical parameters of the cores contaminated by different drilling fluid systems.

Core number	K_1 (mD)	K_2 (mD)	R (%)	Contamination depth (cm)
1	47.45	39.75	83.77	1.1
2	51.56	45.33	87.92	0.9

Notes: 1 is the contamination of the original drilling fluid system; 2 is the contamination of the drilling fluid system with the addition of FCS-08; K_1 is the initial permeability of the core; K_2 is the permeability after the core contamination; R is the return permeability of the core.

tension meter. The experimental temperature was 25°C. The experimental results are shown in Table 4.2.

It can be seen from Table 4.1 that under the same dosage condition (0.4%), FCS-08 showed superior performance in reducing the surface tension compared to other surfactant-type antiwater blocking agents, and the surface tension of water was reduced to below 20 mN/m. According to the Young-Laplace equation the reduction of the interfacial tension between the gas and water (surface tension of the water phase) could greatly reduce the capillary pressure, thereby reducing the possibility of water-blocking damage. It can be seen that FCS-08 could reduce the probability of water-blocking damage in low-porosity and low-permeability formations by reducing the surface tension of the drilling fluid filtrate.

In addition, the higher the oil-water interfacial tension, the higher the water saturation after the drilling fluid filtrate and other foreign fluids invaded the formation, the lower the permeability of the formation, and the more serious the water-blocking damage. Comparing the interfacial tension results of ABSN and FCS-08 aqueous solutions in Fig. 4.7, it can be seen that FCS-08 reduced the interfacial tension of the oil-water interface better than ABSN did. When the concentration of FCS-08 was 0.2%, the oil-water interfacial

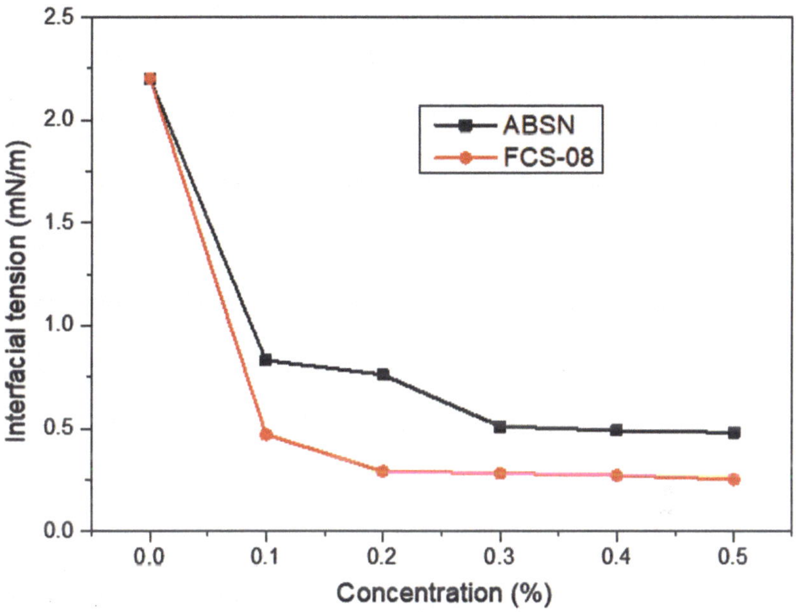

Figure 4.7 Interfacial tension between oil and 0.1%–0.5% ABSN and FCS-08 aqueous solution.

tension could be reduced to 0.32 mN/m, indicating that FCS-08 effectively reduces the oil-water interfacial tension at very low concentrations.

The capillary pressure is a direct linear function of the interfacial tension between the immiscible phases in the porous medium. When the interfacial tension between the fluids decreases, the capillary pressure will decrease. The above experimental results show that FCS-08 effectively reduces the excess capillary pressure of the formation; reduces the power of the foreign liquid to suck into the porous medium, thereby effectively reducing the amount of foreign liquid intrusion near the well zone or fracture surface; and reduces the liquid phase saturation of tight oil and gas formation, increasing the seepage space of oil and gas, greatly improving the seepage conditions of oil and gas near the well zone, and effectively reducing the degree of water blocking.

4.1.2.4 The impact of FCS-08 on the formation

4.1.2.4.1 Capillary rise experiment

In the formation, owing to the existence of many microcapillaries, the liquid-phase fluid in the microcapillary causes the water-blocking effect due to the presence of capillary force, thus causing serious damage to the

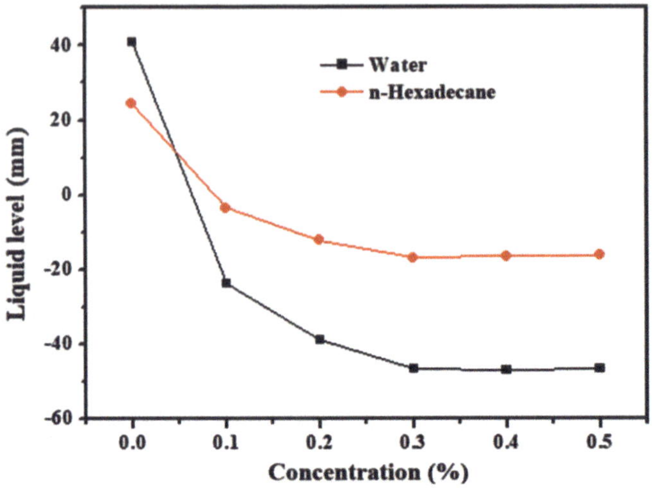

Figure 4.8 Effect of FCS-08 on capillary liquid level.

formation. Through the capillary rise experiment, the influence of FCS-08 on the capillary force was investigated, and the mechanism of protecting the formation was simulated. As shown in Fig. 4.8, after treatment with 0.1% FCS-08, the capillary force of deionized water and n-hexadecane was reduced, and the direction was changed. Analogous to the reversal of the capillary force direction in the formation, FCS-08 transformed the force that originally blocked the pores of the microcapillary into a thrust that promotes the outflow of the liquid phase in the microcapillary, and as the concentration of FCS-08 continues, this effect is more obvious, which proves that FCS-08 solves the water blocking effect existing in the microcapillary of the formation by improving the wettability of the microcapillary surface and achieves the purpose of protecting the formation.

4.1.2.4.2 Dynamic contamination experiment

The physical parameters of the artificial sandstone core before and after contamination of the original drilling fluid system and the drilling fluid with added FCS-08 are shown in Table 4.2. It can be seen that the return permeability after the contamination of original drilling fluid on core was 83.77%, and after adding FCS-08 to the drilling fluid system, the core return permeability after contamination was improved to 87.92%. The core's contamination depth was reduced from 1.1 to 0.9 cm from the original drilling fluid system contamination, further indicating that FCS-08 reduced the core contamination depth and achieved the effect of formation protection.

4.1.3 Field application

Taking the Chang 68 fault block in the Weibei oil and gas field as an example, the design and application of the superamphiphobic drilling and completion fluid system that can protect low-permeability and ultralow-permeability formations are expounded here. There are large differences in the content of clay minerals and the illite-smectite mixed layer in the formation of Chang 68 fault block. The mineral water in the formation is high (about 30,000 mg/L). It is a low-porosity ultralow-permeability formation with a small pore size (the largest pore throat size is 0.494−1.4835 μm), which is a moderately strong water-sensitive and acid-sensitive, medium-salt-sensitive formation with a critical salinity of 30,000 mg/L. The potential damage factors of the formation mainly include clay's hydration expansion and dispersion, water blocking, and inorganic scale. Therefore in designing drilling and completion fluids, a dense cake must be formed on the borehole to effectively block the drilling and completion fluid filtrate from entering the formation, and the fluid invading the formation should have a low interfacial tension to avoid water blocking damage, which is conducive to reverse drainage to reduce formation damage. To this end, through the improvement of the upper drilling fluid, the drilling and completion fluid system with the super-FCS-08 that protect the low-permeability and low-permeability formation were designed.

In the field construction process, in drilling close to the top of the formation, 0.3% FCS-08 was added to modify the upper drilling fluid into the superamphiphobic drilling and completion fluid systems that can protect the low-permeability and ultralow-permeability formations. The system performance was tested, as shown in Table 4.3.

The rheological properties of the drilling fluid before and after the modification did not change much, and the compatibility of FCS-08 with other agents was good. The filtration loss and shale expansion rate of the modified system were reduced, the cuttings recovery and return permeability were greatly improved, and the daily production after commissioning was 1.6 times that of the adjacent well, realizing the protection of low-permeability and ultralow-permeability formations.

Table 4.3 General performance test.

State	Density (g/cm^3)	AV ($mPa \cdot s$)	PV ($mPa \cdot s$)	YP (Pa)	YP/PV	Gel 10″/10′ (Pa)	FL_{API} (mL)	Coefficient of frictional resistance	pH	Cuttings recovery (%)	Shale expansion rate (%)	Return permeability (%)
Before modification	1.10	28	20	8	0.4	1.5/3.0	4.5	0.1	8.5	68.7	5.86	62.33
After modification	1.10	30	20	10	0.5	2.0/3.0	4.0	0.1	8.5	86.2	2.69	93.4

Table 4.4 Comparison of protection effects of low-permeability and low-permeability formations in Shengli Oilfield.

Well number	Protection measures	Daily production (t/d)	Daily oil production per meter in unit pressure differential (t/m · d · MPa)	Production increase by the proposed method
Chun64-4	Proposed method	3.2	0.064	1.408
Chun64-3	Traditional method	0.9	0.045	
Fan143-xie10	Proposed method	5.2	0.168	1.48
Fan143-13	Traditional method	4.7	0.113	
Liang38-ping4	Proposed method	24.8	0.016	1.186
Liang38-Ping2	Traditional method	7.3	0.013	
Liang23-xie36	Proposed method	5.9	0.12	21.83
Liang23-12	Traditional method	0.5	0.005	
Liang38-ping7	Proposed method	10.4	0.012	1.104
Liang38-ping3	Traditional method	2.2	0.011	
Fan147-9	Proposed method	5	0.303	1.697
Fan147-1	Traditional method	2	0.179	

In addition, the system was promoted and applied to other wells in Shengli Oilfield. No complex conditions related to drilling fluids appeared in these wells. Compared with conventional technologies, the damage to formations was significantly reduced. The daily oil production per meter in unit pressure differential greatly increased in comparison with the previous protection technology (Table 4.4), and the formation protection effect and economic benefit were obvious.

The comparison results of field tests (Table 4.5) show that under the same conditions, the protection effect of the superamphiphobic drilling fluid on the formation was far better than that of the other latest technologies in China.

Table 4.5 Comparison of field test results of superamphiphobic temporary plugging and other technologies (ultralow permeability formations).

Well number	Protection measures	Daily fluid (t/d)	Daily oil (t/d)	Water content (%)	Daily oil production per meter in unit pressure differential (t/m · d · MPa)	Monthly decline rate (%)
Gao424-1	Proposed method	4.5	4.4	3	0.2619	−13
Gao 424	Other latest technologies in China	7.4	4.6	28	0.2428	8
Gao 424-Ping1		1.1	1	9	0.0158	6
Chun48-Xie17	Proposed method	9.1	8.3	10	0.2311	3
Chun48-Xie11	Other latest technologies in China	13	0.5	96	0.0096	9
Chun48-13		3.3	3.3	0	0.0023	6
Liang20-Xie31	Proposed method	9.9	4.6	44	2.1296	10
Liang20-17	Other latest technologies in China	12	11	8	1.5151	14
Liang20-Xie21		2.9	0.5	84	0.1428	19

4.2 New technology of bionic drilling fluids to prevent medium-permeability formation damage by imitating the surface structure and characteristics of lotus leaves

4.2.1 Superhydrophobic effect of lotus leaves and its bionic application in protecting formations

The surface of lotus leaves has a strong anticontamination ability. The contaminants falling on their surface can be wrapped and taken away together by the falling water droplets, and the water droplets will not adhere to the surface [88]. This self-cleaning property is called the lotus effect. The mystery is jointly determined by the microstructure papillae on the rough surface of lotus leaves, the branch of the nano-level structure, the hydrophobic waxy substance on the surface, and the concave part between the unique micro-nano composite structure filled with air. The contact between the droplets and the surface of lotus leaves is a solid-gas and liquid-gas composite contact. The droplets can make only a few points of contact with the papillae on the leaf surface, just like sitting on the air cushion. On the other hand, the droplets form a spherical shape under the effect of their own surface tension. Together with the joint support of the waxy layer, air cushion, and papillae, the droplets cannot penetrate the surface but roll freely [88]. The dirt is washed away from the surface as the droplets roll. The self-cleaning properties of lotus leaves' surface provide inspiration for the development of superhydrophobic self-cleaning surfaces. For a solid surface to have superhydrophobic self-cleaning properties, two conditions must be met. First, the surface must have a low surface free energy; second, the surface must have a rough micro-nano geometric structure [88].

During the drilling process, owing to the existence of a positive pressure differential, an outer cake will be formed on the rock of the borehole, and an inner cake will be formed near the borehole by the drilling fluid. Usually, the outer cake is dense and has very low permeability, which is nanostructured. Similar to the microstructure papillae on the surface of lotus leaves, if the microstructure papillae are further enhanced and the surface energy of the cake is reduced, a superhydrophobic surface can be prepared, which not only prevents the solid phase particles in the drilling fluid from entering the formation, but also makes the water contact angle of the cake greater than 90 degrees and cos $\theta < 0$, which prevents the water phase from entering the formation and protects the formation [88]. To achieve this goal, the lotus effect was used to develop a superhydrophobic drilling fluid agent to form a bionic film on the borehole, reduce surface energy, and build a dual structure of micron and nanometers with a certain roughness, which can give it a self-cleaning ability and enables it to protect formations.

4.2.2 Establishment of the new technology of bionic film-type drilling fluid for protecting medium-permeability formations

For medium-permeability formations the second generation of formation protection technology established by Professor Guancheng Jiang of China University of Petroleum (Beijing), that is, formation protection technology based on oil film method, has the disadvantages of a high fluorescence level and difficulty in adapting to the needs of exploring wells. To solve this problem, Professor Guancheng Jiang further introduced bionic films into the field of drilling fluids that can protect formations and developed a film-type amphiphilic polymer formation protector LCM-8, also known as a superhydrophobic agent. By forming a bionic film on the borehole, the solid particles in the drilling fluid are prevented from entering the formation, while ions are allowed to pass through the film under certain conditions. At the same time, it can prevent the water molecules in the drilling fluid from entering the formation by increasing the salinity of the drilling fluid to protect the formation. The purpose is to form a new technology of bionic film-type drilling fluid that protects middle-permeability formations [33].

4.2.2.1 Preparation of LCM-8

Polymers can be prepared by directly connecting the hydrophilic chain and the lipophilic chain by chemical bonds through blocking and grafting and so on. The reactive polymer emulsifier used in this experiment reacted with the water-soluble initiator to form a polymer emulsifier radical, which attacked the latex particle group and reacted according to the micelle nucleation mechanism, as illustrated in Fig. 4.9. The monomer and the reactive polymer emulsifier were added to 150 mL of deionized water for shear emulsification at a certain ratio, and the mixture was transferred to a four-necked flask equipped with a nitrogen-passing device and reflux condensation. The water bath was set at 80°C. The stirrer was started, and the water-soluble initiator was added dropwise to the system in two portions. The reaction was terminated after 6 h. The obtained emulsion was added to the macro and micro bridging agents according to

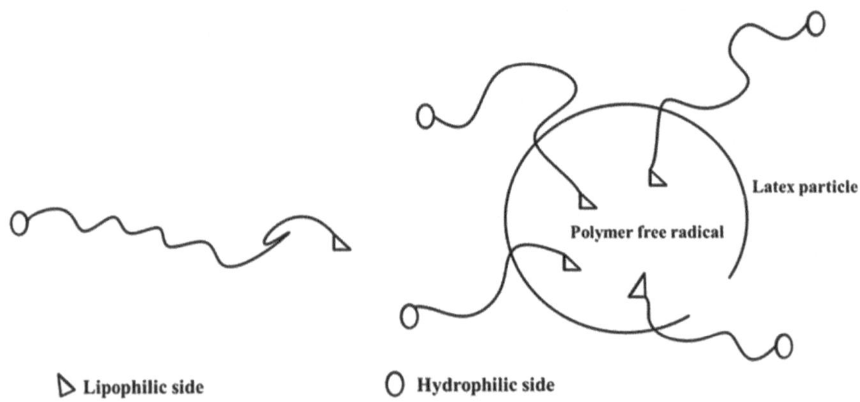

Figure 4.9 Schematic diagram of the mechanism for synthesizing LCM-8.

Figure 4.10 Amphiphilic polymer LCM-8.

a certain ratio, and finally a milky white viscous flowing liquid was obtained, which was LCM-8 (Fig. 4.10).

4.2.2.2 Characterization and performance evaluation of LCM-8

4.2.2.2.1 Particle size analysis

The particle size distribution of LCM-8 is shown in Fig. 4.11. The median particle size of LCM-8 is 0.31 μm, the volume average particle size is 0.33 μm, and the area average particle size is 0.27 μm.

4.2.2.2.2 TEM analysis

LCM-8 was observed by using a high-resolution transmission electron microscopy (TEM) (JEM-2100 LaB6, JEOL, Japan), as shown in Fig. 4.12. It can be seen that the particle size of LCM-8 is less than 100 nm, some particles have a size of less than 50 nm, and the rest have a particle size between 50 and 100 nm. The surface of LCM-8 particles has many rough structures, and there are interconnected branches between the particles. These special physical structures play a vital role in developing the special superhydrophobic properties.

4.2.2.2.3 SEM analysis

A SEM analysis was performed, using an FESEM (Quanta200F, FEI, Netherlands) to detect the micromorphology of the core surface before and after treatment with the superhydrophobic agent. The results are shown in Fig. 4.13. The surface of the untreated core is very smooth on the nano-micro scale, there is only a small amount of layered structure, and it has a very small surface roughness. Therefore the original core surface shows strong hydrophilic wetting characteristics. After the core surface was treated by the superhydrophobic agent, the core surface was covered with many nano-level small particles, and the small particles are connected to each other, thus

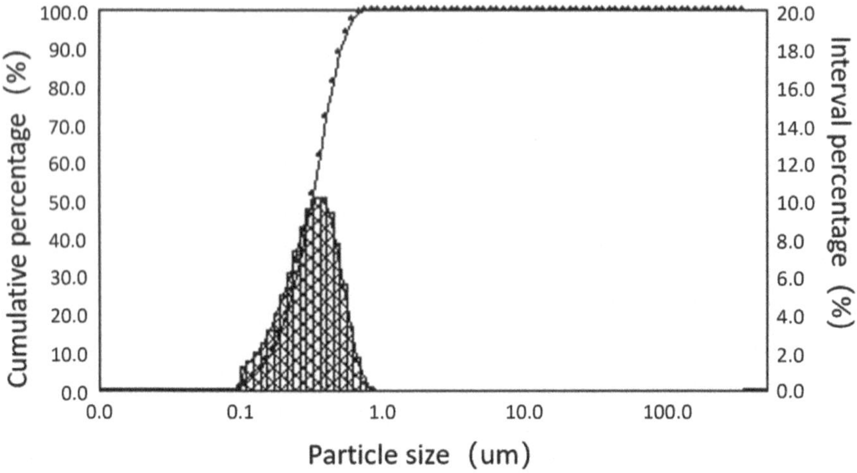

Figure 4.11 Particle size distribution of LCM-8.

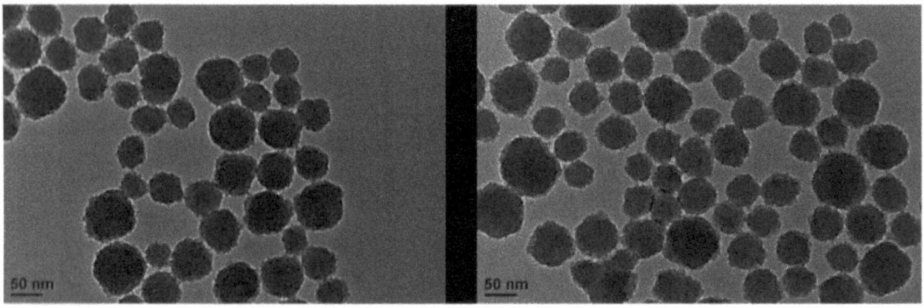

Figure 4.12 TEM images of superhydrophobic agent LCM-8.

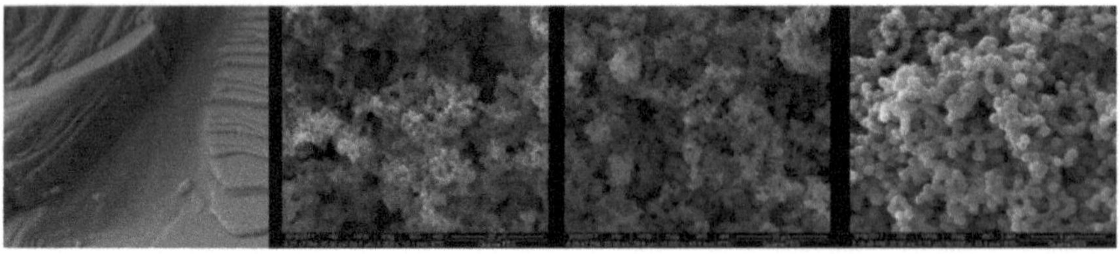

Figure 4.13 SEM images of the core surface before and after treatment with the superhydrophobic agent.

presenting a micrometer-scale physical structure as a whole and further forms a nano-micro structure due to the nano-structure itself. The nano-micro multi-scale physical structure can capture a large amount of air in the nano-scale space and isolate it from the liquid phase, so the treated core surface has super-hydrophobic properties.

4.2.2.2.4 Effect of LCM-8 on the wettability of the core surface

1. Measurement of the contact angle on the core surface

The synthesized superhydrophobic agent was used to prepare solutions with different concentrations. The core pieces were immersed in the solution for 4 h and then taken out and dried in a constant-temperature drying oven at 80°C to constant mass. Then a contact angle measuring instrument (JC2000D3, Shanghai Zhongchen Digital Technic Apparatus, China) was used to measure the water contact angle on the artificial core surface treated with a superhydrophobic agent. The results are shown in Fig. 4.14. The water contact angle on the original core surface was 34 degrees, and when the core was treated with the superhydrophobic agent, the contact angle increased continuously with the increase in the concentration of the superhydrophobic agent and finally remained constant. When the mass concentration was 0.1%, the contact angle increased to 85 degrees. When the concentration increased to 0.3%, the contact angle increased to 150 degrees. The concentration continued to increase, and the contact angle remained basically unchanged at 150 degrees, indicating that it had turned from hydrophilic to hydrophobic.

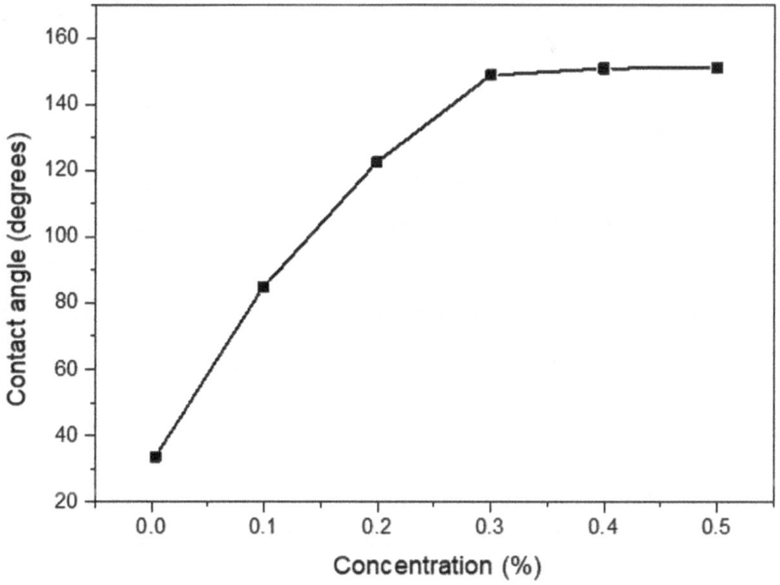

Figure 4.14 Contact angle of the core surface after treatment with different concentrations of the superhydrophobic agent.

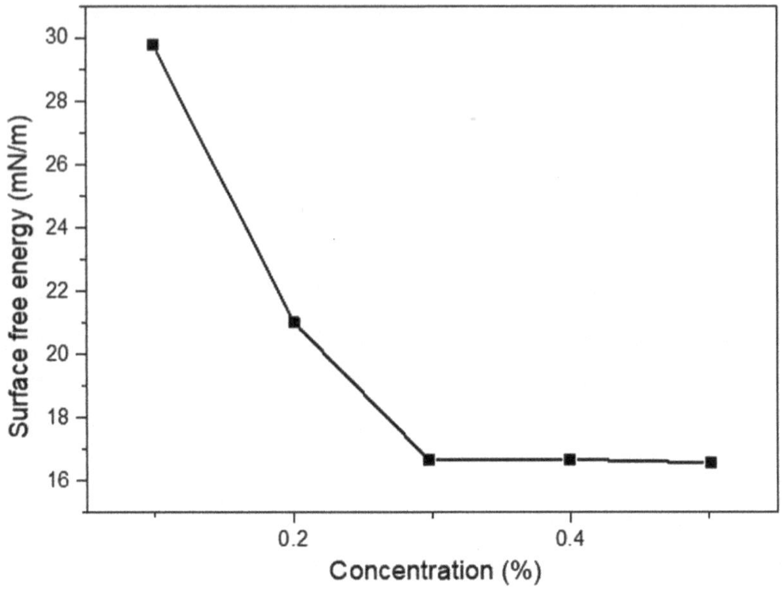

Figure 4.15 Free energy change of the core surface after treatment with the superhydrophobic agent.

2. Calculation of free energy on the core surface

   Using the Owens two-liquid method and the measured water and oil contact angles, the surface free energy of the artificial core after treatment with the superhydrophobic agent was calculated. The results are shown in Fig. 4.15. The surface free energy of the original core was 62 mN/m. When the core surface was treated with the superhydrophobic agent, the core surface free energy decreased continuously as the concentration increased. When the concentration reached 0.3%, the core surface free energy (surface tension) decreased to below 17 mN/m and remained stable. Therefore the superhydrophobic agent effectively reduces the core surface free energy.

4.2.2.2.5 Evaluation of the shale inhibition of the superhydrophobic agent

1. Linear expansion experiment

   Through the linear expansion experiment of bentonite in superhydrophobic solutions of different concentrations, the shale inhibitory performance of the superhydrophobic agent was investigated. The results are shown in Fig. 4.16. With the amount of superhydrophobic agent increasing, the linear expansion height of bentonite was significantly

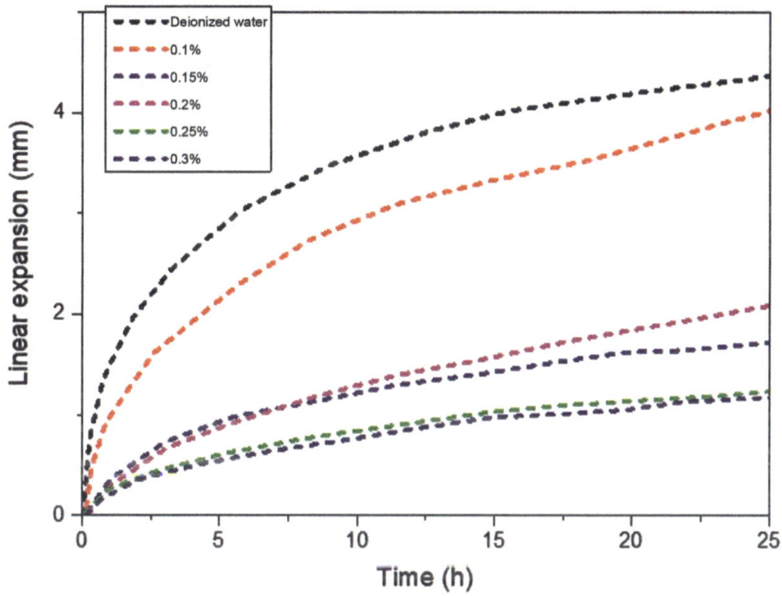

Figure 4.16 Linear expansion height of bentonite in different concentrations of superhydrophobic agent solution.

Table 4.6 Rolling recovery of shales in different concentrations of superhydrophobic agent solutions (hot rolled at 120° for 16 h).

Sample information	Rolling recovery (%)
Shale + pure water	14
Shale + 0.1% superhydrophobic agent	33
Shale + 0.2% superhydrophobic agent	52
Shale + 0.3% superhydrophobic agent	78
Shale + 0.4% superhydrophobic agent	77

reduced. After 24 h of immersion it was found that when the amount of superhydrophobic agent was 0.3%, the linear expansion of bentonite decreased by about 50% relative to that immersed in pure water.

2. Hot rolling recovery experiment

The results of the rolling recovery experiment of shale in pure water and different concentrations of superhydrophobic agent are shown in Table 4.6. At 120°C, with the content of the superhydrophobic agent increasing, the rolling recovery of the cuttings increased continuously. When the superhydrophobic agent content reached 0.3%, the rolling recovery of cuttings was as high as 78%. The results were consistent with those of the linear expansion experiment, further demonstrating that the superhydrophobic agent could not only inhibit the hydration and expansion of shale, but also successfully inhibit the dispersion and exfoliating of shale.

4.2.2.2.6 Evaluation of the superhydrophobic agent on formation protection performance

1. Experiment of natural imbibition of the core before and after treatment with the superhydrophobic agent

   The natural imbibition experiment used a self-designed core natural imbibition measuring device to determine the imbibition content of the artificial core to the water phase before and after treatment with the superhydrophobic agent. The test liquid for the water phase was deionized water. The test samples were prepared as follows: First, the core was placed in a 105°C oven and dried for 4 h. Then the core was immersed in a certain amount of superhydrophobic agent solution for 12 h. Next, the core was placed in a 105°C oven, dried to constant mass, and taken out for determining the natural imbibition content of the core. The test results are shown in Fig. 4.17. When the core was not treated with the superhydrophobic agent, the core's natural imbibition content of

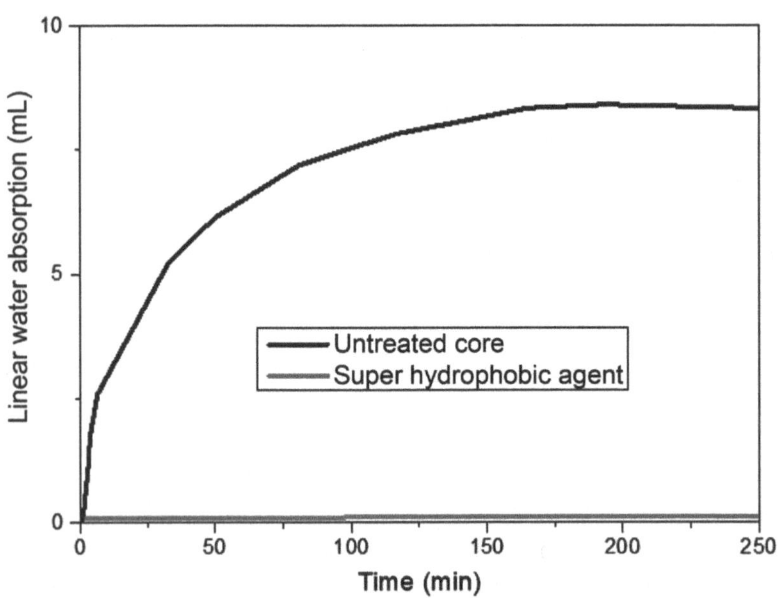

Figure 4.17 Schematic diagram of core's natural imbibition.

Table 4.7 Changes in physical parameters of the cores contaminated by different drilling fluid systems.

Core number	K_1 (mD)	K_2 (mD)	R (%)	Contamination depth (cm)
1	9.45	7.70	81.48	0.95
2	11.23	10.47	92.23	0.50

Notes: 1 is the core contaminated by the original drilling fluid system; 2 is the core contaminated by the drilling fluid system with the addition of superhydrophobic agent; K_1 is the initial permeability of the core; K_2 is the permeability after the core contamination; R is the return permeability of the core.

deionized water reached 8.38 mL. After being treated with the superhydrophobic agent, the value was only 0.025 mL, indicating that after the core was treated with the superhydrophobic agent, a hydrophobic film was formed on the surface of the core pore throat, so the wettability of the pore throat surface inside the core changed from hydrophilic to hydrophobic, preventing fluid from entering the pore throat.

2. Dynamic contamination experiment

   The physical parameters of the artificial sandstone core before and after the contamination by the original drilling fluid system and the one added with superhydrophobic agent are shown in Table 4.7. The return permeability of the core contaminated by the original drilling fluid was 81.48%, and after adding superhydrophobic agent to the drilling fluid system, the core return permeability after contamination was improved to 92.23%. The contamination depth of the core was reduced from 0.95 cm in the original drilling fluid system to 0.50 cm, which further showed that the superhydrophobic agent achieved the effect of formation protection.

3. Characterization of core surface performance after dynamic contamination

   The contact angle test and SEM analysis were performed on the cores after dynamic contamination. When the core was contaminated with the original drilling fluid system and with the drilling fluid system to which the superhydrophobic agent had been added, the contact angle of the core was 39.18 and 128.87 degrees, respectively (Fig. 4.18). In addition, from Fig. 4.19 we see that the surface of the core contaminated by the original drilling fluid system was covered with a thick layer of soil but there were still many pore throats on the surface, which was conducive to water imbibition; for surface of the core contaminated by the drilling fluid system to which the superhydrophobic agent had been added, there were many small hydrophobic particles (superhydrophobic

Figure 4.18 Measurement of the contact angle of the core surface after contamination. The original drilling fluid system (*left*). The drilling fluid system with the addition of the superhydrophobic agent (*right*).

Figure 4.19 SEM images of the core surface after contamination. The original drilling fluid system (*left*). The drilling fluid system with the superhydrophobic agent (*right*).

agent), which gave the core surface strong hydrophobic properties, thus preventing water from entering the core and protecting the formation.

4.2.2.2.7 Effect of the superhydrophobic agent on the performance of the drilling fluid system

The effect of the superhydrophobic agent on the rheological properties and filtration loss of the drilling fluid system was investigated. The results are shown in Table 4.8. After the superhydrophobic agent was added, there was no difference in the rheological properties of the system at room temperature. After hot rolling at 120°C, the apparent viscosity (AV) and plastic viscosity (PV) of the system were reduced to a certain extent. The medium-pressure filtration loss of the agent system was reduced by 40%, and the high temperature and high pressure (HTHP) filtration loss was reduced by 30%, indicating that the superhydrophobic agent could reduce the filtration loss after high temperature.

Table 4.8 Effect of superhydrophobic agent on the performance of drilling fluid system.

Temperature	Superhydrophobic agent concentration (wt.%)	AV (cP)	PV (cP)	YP ($N \cdot m^{-2}$)	YP/PV	G10″/G10′ (Pa)	FL_{API} (mL)	FL_{HTHP} (mL)
20°	0	57.5	46	11.5	0.25	2/13		
	0.3	57.5	43	14.5	0.34	2.5/14		
120°	0	67.5	51	16.5	0.32	5/19.5	6.6	14.0
	0.3	61.0	45	16	0.36	3.5/15.5	3.8	10.2

HTHP, high temperature and high pressure.

4.2.2.2.8 Analysis of film-forming performance

SEM analysis was performed on the filter paper's surface, the filter paper's side surface, and the sandstone surface before and after filming. The results are shown in Figs. 4.20–4.22. As can be seen, LCM-8 has good film adhesion on filter paper or on the surface of sand and gravel.

4.2.2.2.9 Comparison of LCM-8 with similar products

Sodium bentonite was used to prepare a base mud with a density of 1.03 g cm^3. Various types of filming agents were added to the base mud to determine the rheological properties and API filtration loss as well as the HPHT filtration loss of the drilling fluid under the conditions of 80°C, 3.5 MPa, and 30 min. The results are shown in Tables 4.9 and 4.10.

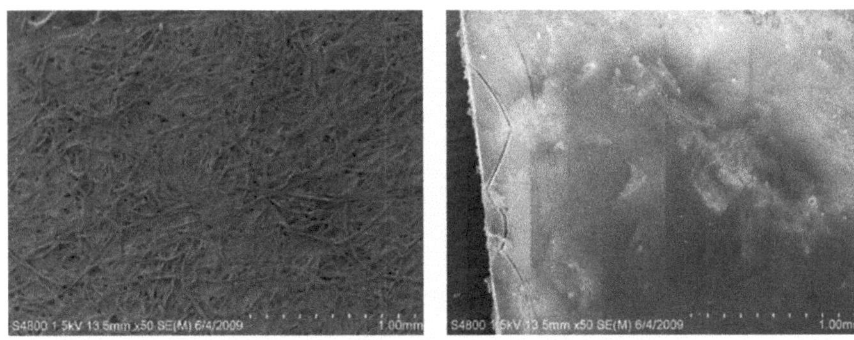

(A) (B)

Figure 4.20 SEM images of the filter paper's surface before and after the filming by LCM-8. (A) SEM images of the filter paper's surface. (B) LCM-8 filmed on the surface of the filter paper.

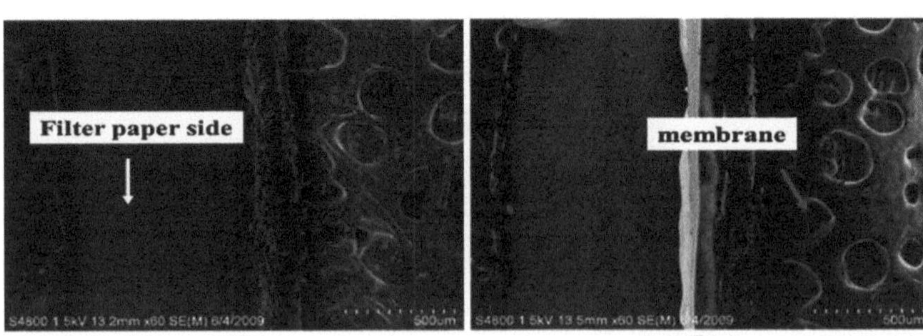

(A) (B)

Figure 4.21 SEM images of the filter paper's side surface before and after the filming by LCM-8. (A) SEM images of the filter paper's side face. (B) LCM-8 filmed on the side face of the filter paper.

(A) (B)

Figure 4.22 SEM images of the filter paper's sand surface before and after the filming by LCM-8. (A) Sand surface. (B) LCM-8 filming on the sand surface.

Table 4.9 Comparison of HTHP filtration loss of each filming agent under different dosages (mL).

Added amount (%)	0	1	2	3	4	5	6
FLC2000 (international advanced product)	90	64	56	44	37	32	25
LCM-8	90	48.8	40.2	34.6	28.4	23	21.2
Chinese advanced product 1	90	81.8	64.6	54.2	38.4	35.4	26.2
Chinese advanced product 2	90	88	84	80.4	85	87	/

HTHP, high temperature and high pressure.

Under different dosage conditions, LCM-8 had almost no effect on the rheological properties of the base mud and exhibited a better effect on reducing API filtration loss and HPHT filtration loss than other film materials, indicating that LCM-8 had better filming ability and stronger plugging ability.

4.2.2.3 Establishment of the new technology of bionic film-type drilling fluids for protecting medium-permeability formations

The amphiphilic polymer agent prepared according to the bionics and polymer chemistry theories has a strong ability to form molecular ordered bodies. The agent can be attached to the surface of filter paper, cake, and rock at a certain temperature and pressure. At the same time, the polymer has oil-soluble and water-soluble amphiphilic properties. When added to water-based drilling fluid, it can be quickly adsorbed on the rock surface of the borehole. When the critical concentration is reached, the polymer

Table 4.10 Comparison of the rheological properties and API filtration loss of each filming agent under different dosages.

Type	Added amount (%)	AV (mPa · s)	PV (mPa · s)	YP (Pa)	YP/PV (Pa/(mPa · s))	API. V_f (mL)
Chinese advanced product 1	0	10	5	5	1	34
	1	10	4.5	5.5	1.22	30
	2	17.5	8.5	9	1.06	26.4
	3	19.75	9.5	10.75	1.13	23.8
	4	22.5	8	14.5	1.81	20.4
	5	32	13.5	18.5	1.37	16.8
	6	35	18	17	0.94	14.2
Chinese advanced product 3	0	10	5	5	1	34
	1	9	4	5	1.25	32.8
	2	12.75	7.5	5.25	0.7	30.8
	3	16.25	8.5	7.75	0.91	31
	4	18.5	9.5	9	0.95	27.6
	5	21.5	11.5	10	0.87	26.4
	6	24	12	12	1	27
Chinese advanced product 2	0	10	5	5	1	34
	1	11	6	5	0.83	20.4
	2	13.25	7.5	5.75	0.77	17.4
	3	12.5	7	5.5	0.78	14.2
	4	13	6.5	6.5	1	11.6
	5	14	7.5	6.5	0.87	8.6
	6	17.5	8.5	9	1.06	8
LCM-8	0	10	5	5	1	34
	1	16.5	13.5	3	0.22	12
	2	38	27	11	0.41	8
	3	42.5	28	14.5	0.52	7.4
	4	55	34	21	0.62	6.8
	5	70	39	31	0.79	6.4
	6	86.5	41	45.5	1.11	5.6

associates on the rock surface to form a hydrophobic microregion. The hydrophobic microregion has a hydrophobic core composed of the hydrophobic groups of the polymer, and the outer layer is surrounded by the hydrophilic segment of the polymer.

As the polymer concentration increases, a large number of micelles of different sizes are formed on the borehole, and the polymer develops from intrachain association to interchain association, thereby forming a plugged zone on the rock surface, which makes the drilling fluid and its filtrate almost completely isolated and unable to penetrate the formation. It can realize near-zero filtration loss, which effectively prevents the migration of particles in the formation, reduces the damage of drilling fluid to the formation, and has a significant protective effect.

The film-type drilling fluid not only can effectively plug and protect the medium-permeability formation, but also can expand the safety density window of the drilling fluid, which can better solve the technical problems such as lost circulation, pipe sticking, collapse, and formation damage in drilling long-open-hole multiple pressure formations or pressure failure formations.

In summary, the new technology of bionic film-type drilling fluid with ultralow formation damage suitable for medium-permeability formations is upper drilling fluid + 2%–3% LCM-8 for formation protection.

4.2.3 Comparison of the effectiveness of the new technology of bionic film-type drilling fluids and traditional technologies for protecting medium-permeability formations

4.2.3.1 Comparison of indoor evaluation data

A comparison of experimental data of the visual sand bed is shown in Fig. 4.23. The penetration depth of LCM-8 is lower than that of the compared samples, indicating that LCM-8 can form a higher-quality film on the sand bed.

The performance of LCM-8 in the base mud was significantly better than that of other film materials (including the latest international products), indicating that LCM-8 has better filming ability and stronger plugging ability.

In terms of the effect of formation protection, the plugging rate refers to the ability to form a shielding temporary plugged zone with a seepage capacity of almost zero, and the recovery of seepage capability refers to the ability to remove the plugged zone. Obviously, the higher the plugging rate and the

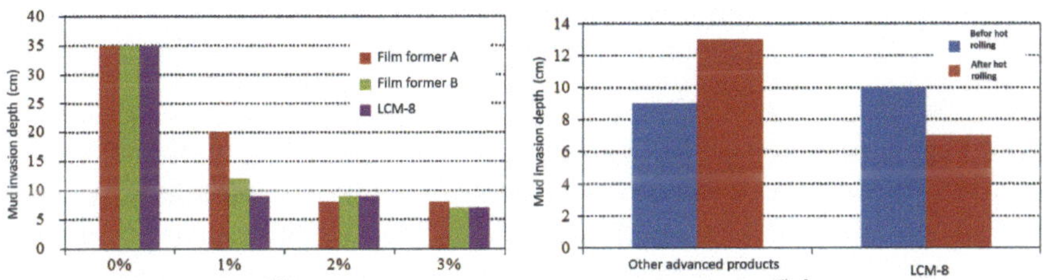

Figure 4.23 Comparison of the penetration depth of several filming agents (additional amount: 2%).

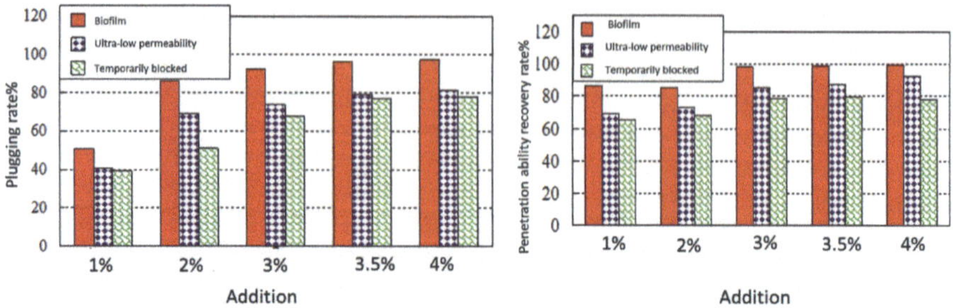

Figure 4.24 Indoor effect comparison of the ultralow-damage bionic film-type temporary plugging technology and ultralow-permeability technologies.

recovery of seepage capacity, the better the protection effect on the formation. It can be seen from Fig. 4.24 that the newly developed ultralow-damage bionic film-type temporary plugging technology is far better than other technologies in protecting formations.

4.2.3.2 Comparison of field application effects

Field tests show that, in comparison with similar wells, the skin coefficient of the well using ultralow-damage bionic film-type temporary plugging technology was reduced by more than 99% over the previous technologies, achieving the goal of ultralow damage. Under the same conditions the oil production index per meter increased by 54%–103%, and the apparent water absorption index per meter on the well increased by 139%–750%. In Dagang Oilfield, 30% of the test wells achieved spontaneous oil recovery, which had never been seen before. The field test data of some wells are shown in Tables 4.11 and 4.12. The oil production index per meter refers to the daily oil production per unit of formation under the unit pressure differential; the apparent water absorption index per meter refers to the daily water injection per unit of formation under the unit pressure differential; and the skin coefficient describes the degree of damage to the formation. The lower the skin coefficient, the smaller the damage, and when the skin coefficient is zero, it means that the formation is not damaged (sandstone formation).

Whether for a water well or an oil well, the field test results show that with the application of ultralow-damage bionic film-type temporary plugging technology, the skin coefficient was very low, and the oil production and water injection were increased by many times, indicating that the use of this technology had achieved ultralow damage to the formation and has greatly improved the formation protection technology.

Table 4.11 Comparison of field testing results of ultralow-damage bionic film-type temporary plugging technology and other technologies on oil wells (medium-permeability formations).

Well type	Well number	Drilling fluid system	Skin coefficient	Oil production index per meter (m^3/(d · m · MPa))	Increase rate of oil production index per meter (%)
Test well	Guan27-55	Ultralow-damage bionic film-type temporary plugging	0.01	0.048	71.43/54.84
Compared well	25-51	Other formation protection technologies	3.55	0.031	–
Compared well	Guan25-53	Other formation protection technologies	3.26	0.028	–
Test well	Gang393-1	Ultralow-damage bionic film-type temporary plugging	0.01	0.81	42.11
Compared well	Gangshen41 × 1	Other formation protection technologies	3.41	0.51	–
Test well	Gang393-1	Ultralow-damage bionic film-type temporary plugging	0.02	6.18	63.49
Compared well	Gang26-8	Other formation protection technologies	3.45	3.78	–
Test well	Zao76-13	Ultralow-damage bionic film-type temporary plugging	0.02	0.634	103.21
Compared well	Zao74-13	Other formation protection technologies	5.33	0.312	–

Table 4.12 Comparison of field testing results of ultralow-damage bionic film-type temporary plugging technology and other technologies on water wells (medium-permeability formations).

Well type	Well number	Drilling fluid system	Skin coefficient	Apparent water absorption index per meter ($m^3/(d \cdot m \cdot MPa)$)	Increase rate of apparent water absorption index per meter (%)
Test well	Guan23-52	Ultralow-damage bionic film-type temporary plugging	0.01	0.46	751.85
Compared well	Guan27-59	Other formation protection technologies	4.62	0.054	–
Test well	Gang49-8	Ultralow-damage bionic film-type temporary plugging	0.03	1.11	139.22
Compared well	Gang49-2	Other formation protection technologies	3.91	0.464	–
Test well	Wang45-9	Ultralow-damage bionic film-type temporary plugging	0.04	0.62	292.41
Compared well	Wang41-3	Other formation protection technologies	5.12	0.158	–

4.3 New technology of bionic synergistic drilling fluids to prevent high-permeability formation damage by imitating the multilayer composite structure of shells

4.3.1 Shells' structural characteristics and their bionic application in protecting formations

The main components of shells are 95% calcium carbonate and a small amount of chitin. Generally, shells have three layers. The outermost layer is dark brown cuticle (shell skin), the middle layer is the prismatic layer (shell layer), and the inner layer is the nacre layer (bottom layer). The organic layer and mineral layer in the nacre layer are arranged alternately, and the brick-mud structure can be observed on close inspection. This structure is the key to the shell's toughness. In the process of nacre deformation and fracture, the organic matrix and the adjacent aragonite layer adhere to each other, forming a multilevel brick-mud structure. There is a strong interface between the organic phase and the aragonite flakes, which increases the slip resistance between the adjacent aragonite layers and enhances the toughening effect of fiber pull-out. On the other hand, the organic matrix is connected to the separate aragonite layers like a bridge, which reduces the stress field intensity factor at the crack tip and enhances the crack propagation resistance, thereby improving the toughness of the material [39]. Inspired by the special structure of shells, researchers have sought to use the method to design bionic materials. By exploring the relationship between structure and function and combined with experimental characterization methods, the performance parameters of shells are measured, the laws are summarized, and the structure and operation mechanism of shells are revealed. This research has gone deep into bionics, using bionic design methods and concepts to develop new lightweight, high-strength, supertough layered composite materials to solve technical problems that cannot be solved by conventional technical means, such as the ideal effect of formation protection [89].

For high- and ultrahigh-permeability formations, the pore throat diameter of the formation is relatively large. The main mechanism of formation damage is that the solid-phase particles and liquid-phase damage are equally important. The pressure differential easily causes damage to the film. The technology of changing the surface properties of the rock also makes it difficult to solve the problem of clogging of solid particles. Therefore it is necessary to establish corresponding protection technologies for the damage characteristics of high- and ultrahigh-permeability formations. Inspired by the shells' special multilevel brick-mud structure to form a high level of toughness, the team led by Professor Guancheng Jiang developed a bionic high-strength supertough layered composite drilling fluid material composed of organic matter and inorganic particles that can protect high-permeability formations. The organic matter and inorganic particles are served by LCM-8 and the ideal filler, respectively, and the particle size of the ideal filler is determined by the pore throat size of the formation and calculated according to the ideal filling theory. This technology uses the brick-mud alternating multilayer composite structure as a model. It combines film technology with ideal filling technology. The ideal filling technology is used to change the large hole throats into small ones on which films are formed to improve film quality. Thus, through the synergistic effect between the two, the target of ultralow damage to high- and ultrahigh-permeability formations is achieved [90,91].

4.3.2 The establishment and performance evaluation of the new technology of bionic synergistic drilling fluid to protect high-permeability formations

For high and ultrahigh-permeability formations, the sole use of a new generation of filming agent LCM-8 cannot achieve a good plugging effect. To quickly plug the large pores and throats of high-permeability formations, synergistic effect theory is proposed. Through the invention of a new generation of filming agent and ideal filling agent, the formation protection technology with the synthetic effect of film-type drilling fluid and ideal filling is established. As a result, the goal of rapid plugging and filming, zero intrusion, and ultralow damage to high and ultrahigh permeability formations has been achieved (Fig. 4.25).

The research results show that the new technology of ultralow-damage synergistic drilling fluid suitable for high and ultrahigh permeability formations is composed of upper drilling fluid + 1%–2% LCM-8 + 2%–3% ideal filling agent. Here, the zwitterionic polymer drilling fluid system is used as an example to explain the performance evaluation of the synergistic drilling fluid technology with formation protection ability.

To form a new system of zwitterionic polymer synergistic drilling fluid for protecting formations, the basic formula of zwitterionic polymer drilling fluid is as follows: 4% Bentonite + 0.15% FA-367 + 0.2% OYW-012 + 1.5% MAC-023 + 1.5% sulfonated lignite + 1% sulfonated phenolic resin + 1% sulfonated asphalt, followed by the addition of 2% LCM-8 and 3% ideal filler.

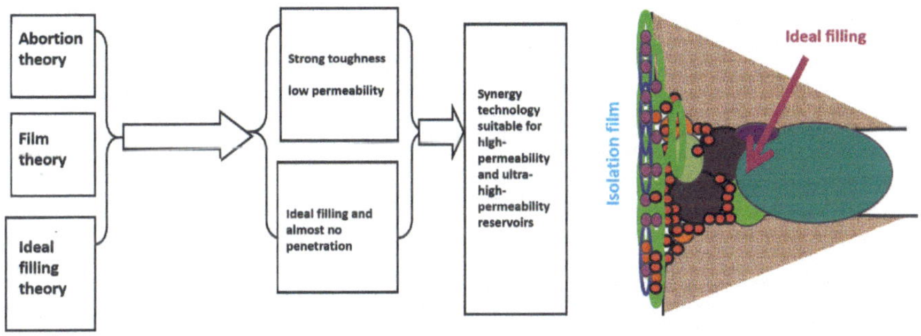

Figure 4.25 The principle of synergistic effect method.

4.3.2.1 Drilling fluid performance evaluation

To investigate the compatibility of the preferred LCM-8 and the designed ideal filling composite temporary plugging agent with the amphoteric composite ion drilling fluid, the changes in performance of the amphoteric composite ion drilling fluid with the addition of the above two agents were evaluated. The evaluation results were compared with the experimental results of other groups. The evaluation experiment results are shown in Table 4.13.

From the experimental results, a preferred drilling fluid system based on base mud + filming agent + ideal filling temporary plugging agent can be obtained. The performance indicators of the system are as follows:

1. The preferred filming agent had good compatibility with the ideal filling temporary plugging agent and with the base mud of amphoteric composite ion drilling fluid.
2. The system had low filtration loss and could play a good role in protecting formations. It can be seen from the experimental results that the API filtration loss was $3.2 < 5.0$ mL; the filtration loss increased to 3.8 mL after hot rolling at 120° for 16 h, and the filtration loss increased to 4.6 mL after hot rolling at 140° for 16 h. The filtration loss was small, which proved that the system had good high-temperature resistance (140°).
3. The rheological parameters of the synergistic drilling and completion fluids with formation protection ability were reasonable, with a filtration loss lower than that when the filming agent is used alone or when the ideal filling temporary plugging agent is used alone, and had the effect of superimposed synergism. It can be seen from the experimental results that the YP/PV ratio at room temperature was 0.6. It was reduced to 0.57 after hot rolling at 120° and reduced to 0.5 after

Table 4.13 Evaluation of drilling fluid performance after adding filming agent and ideal filling temporary plugging agent.

Formula		Condition	ρ (g/cm³)	AV (mPa · s)	PV (mPa · s)	YP (Pa)	Gel (Pa/Pa)	pH	API-V_f (mL)
Basic formula		Room temperature	1.07	30	19	11	1.5/3	9	7.0
		120° × 16 h	1.07	25.5	18	7.5	0.5/4	9	8.2
Basic formula + 2% LCM-8		1.07	38	24	14	2/8	9	5.2	
	120° × 16 h	1.07	36	23	13	2/7	9	6.2	
Basic formula + 4% ideal filling temporary plugging agent	Room temperature	1.08	37.5	24	13.5	2/8	9	5.4	
	120° × 16 h	1.08	35	22	13	2/7	9	5.8	
Basic formula + 2% LCM-8% + 4% ideal filling temporary plugging agent	Room temperature	1.08	42	25	17	2/9	9	3.6	
	120° × 16 h	1.08	44	28	16	2/8	8	4.4	
Basic formula + 2% LCM-8% + 4% ideal filling temporary plugging agent + barite	Room temperature	1.21	48	30	18	3/11	9	3.2	
	120° × 16 h	1.21	47	30	17	3/9	8	3.8	
	140° × 16 h	1.21	43	28	14	2/10	8	4.6	

hot rolling at 140°, and the overall change was not large. This indicates that the system had a good rock-carrying effect. Therefore after optimization the final formula of the synergistic drilling and completion fluid is as follows:

4% Bentonite + 0.15% FA-367 + 0.2% OYW-012 + 1.5% MAC-023 + 1.5% sulfonated lignite + 1% sulfonated phenolic resin + 1% sulfonated pitch + 2% LCM-8 + 4% ideal filling composite temporary plugging agent.

4.3.2.2 Evaluation of dynamic damage

The JHCF-1 core dynamic damage contamination tester was used to conduct dynamic damage experiments on the following experimental formulas. The core of well Xing-27 was selected as the experimental core. The experimental results are shown in Tables 4.14 and 4.15.

1: 4% Soil + 0.15% FA-367 + 0.2% OYW-012 + 1.5% MAC-023 + 1.5% sulfonated lignite + 1% sulfonated phenolic resin + 1% sulfonated pitch

2: Formula 1 + 2% LCM-8 (filming agent)

3: Formula 1 + 4% ideal filling temporary plugging agent

4: Formula 1 + 2% LCM-8 (filming agent) + 4% ideal filling temporary plugging agent

5: The content of barite was increased to 1.25 g/cm^3 based on formula 4.

The results of the dynamic damage experiments show that the addition of a film-filling agent or an ideal filling temporary plugging agent to the optimized zwitterionic polymer drilling fluid is conducive to improving the core plugging rate and the protection effect of formations. However, when these two agents are applied simultaneously to the optimized zwitterionic polymer drilling fluid, a dense cake can be formed near the

Table 4.14 Evaluation results of the dynamic damage of zwitterionic polymer drilling and completion fluid system.

Core sample	Formula	K_{o1} ($\times 10^{-3}$ μm^2)	K_{o2} ($\times 10^{-3}$ μm^2)	K_m ($\times 10^{-3}$ μm^2)	K_{o2}/K_{o1} (%)	$1 - K_m/K_{o1}$ (%)	Starting pressure differential (MPa)
1	1	1.1	0.44	0.27	40	74.8	5.34
2	2	1.3	0.81	0.2	62	84.6	1.35
3	3	1.68	1.36	0.23	81	86.3	1.23
4	4	0.92	0.78	0.05	85	94.6	0.74
5	5	1.23	1.09	0.00984	89	99.2	0.71

Table 4.15 Evaluation of the penetration depth of solid phase for the dynamic damage of zwitterionic system.

Core sample	K_{o1} ($\times 10^{-3}$ μm^2)	Natural kerosene backflow		Cut the cake's end by 1 cm		Cut the cake's end by 2 cm	
		K_{o2} ($\times 10^{-3}$ μm^2)	K_{o2}/K_{o1} (%)	K_{o3} ($\times 10^{-3}$ μm^2)	K_{o3}/K_{o1} (%)	K_{o4} ($\times 10^{-3}$ μm^2)	K_{o4}/K_{o1} (%)
1	1.1	0.44	40	0.56	51	0.68	62
2	1.3	0.81	62	0.936	72	1.053	81
3	1.68	1.36	81	1.39	82.7	1.42	84.5
4	0.92	0.78	85	0.81	88	0.86	93.5
5	1.23	1.09	89	1.22	99	1.22	99

Notes: K_{o3} is the oil permeability measured after cutting the cake by 1 cm; K_{o4} is the oil permeability measured after cutting the cake's end by 2 cm.

Table 4.16 Comparison of field test results of ultralow-damage synergistic temporary plugging and other technologies (high-permeability formations).

Well number	Formation thickness (m)	Daily production (m^3/d)			Oil production index per meter ($m^3/m \cdot MPa$)	Number of comparable well	Formation thickness (m)	Daily production of construction well (m^3/d)			Oil production index per meter ($m^3/m \cdot MPa$)	Production increase rate
		Fluid	Oil	Water content				Fluid	Oil	Water content		
Zhong30-Xiegengg533	10.5	42.1	10	78.8	0.952	Zhong31-Geng533	7.3	27.3	3.3	88.1	0.452	2.11
Zhong31-Xie530	15.2	40.5	13	65.8	0.855	Zhong33-531	16.2	3.3	13.9	88.1	0.858	≈1
Zhong32-Xie533	15	35.3	8.7	75.3	0.58	Zhong31-Xie533	7.3	27.31	3.3	88.1	0.452	1.28
Zhong30-Xiegeng528	19.1	38.7	14	62.2	0.733	Zhong31-Xie529	15.3	10.7	1.3	88.2	0.085	8.63
Zhong31-Pai532	6.8	21.3	7.5	65	1.103	Zhong31-531	12	9.7	1.3	86.4	0.108	10.2
Zhong33-Pai528	10.8	56.7	2.9	94.8	0.269	Zhong33-Geng529	12.1	71.3	2.7	96.2	0.223	1.2
Zhong35-Xie528	10.2	57.6	4.6	92.1	0.451	Zhong33-526	14.1	50.9	4.9	90.4	0.348	1.3

borehole, which significantly reduces the penetration of solid-phase particles into the formation. This further reduces the damage of drilling and completion fluids to the core of the formation, and the permeability recovery of the natural reverse of the core reaches 85%.

4.3.3 Comparison of field application effects

The field test results (Table 4.16) show that under the same conditions, the new technology of ultralow-damage synergistic drilling fluid had a much better protection effect on formations than other Chinese new technologies.

References

[1] J. Zhang, Analysis of the flow of drilling fluid in the formation, Sci. Technol. Get. Rich. Guide (36)(2010) 337−338.
[2] H. Santos, M. Villas-Boas, R. Lomba, et al., API filtrate and drilling fluid invasion: is there any correlation? Proceedings of the Latin American and Caribbean Petroleum Engineering Conference, Society of Petroleum Engineers, 1999.
[3] F. Labenski, P. Reid, H. Santos, Drilling fluids approaches for control of wellbore instability in fractured formations, Proceedings of the SPE/IADC Middle East Drilling Technology Conference and Exhibition, Society of Petroleum Engineers, 2003.
[4] H. Santos, J. Olaya, No-damage drilling: how to achieve this challenging goal? Proceedings of the IADC/SPE Asia Pacific Drilling Technology, Society of Petroleum Engineers, 2002.
[5] M.E. Semmelbeck, J.T. Dewan, S.A. Holditch, Invasion-based method for estimating permeability from logs, SPE Annual Technical Conference and Exhibition, Society of Petroleum Engineers, 1995.
[6] P. Reid, H. Santos, Novel drilling, completion and workover fluids for depleted zones: avoiding losses, formation damage and stuck pipe, Proceedings of the SPE/IADC Middle East Drilling Technology Conference and Exhibition, Society of Petroleum Engineers, 2003.
[7] H. Santos, R. Perez, What have we been doing wrong in wellbore stability? Proceedings of the SPE Latin American and Caribbean Petroleum Engineering Conference, Society of Petroleum Engineers, 2001.
[8] M.A. Dick, T.J. Heinz, C.F. Svoboda, et al., Optimizing the selection of bridging particles for reservoir drilling fluids, SPE International Symposium on Formation Damage Control, Society of Petroleum Engineers, 2000.
[9] N. Hands, K. Kowbel, S. Maikranz, et al., Drill-in fluid reduces formation damage, increases production rates, Oil Gas J. 96 (28) (1998) 65−69.
[10] J. Ye, Discussion on Drilling and Completion Fluid Technology for Protecting Oil and Gas Layer in Deep Exploration Well, West-China Exploration Engineering, 2006.
[11] G.H. Fancher, J.A. Lewis, K.B. Barnes, Some physical characteristics of oil sands. [porosity, permeability, and screen analysis], Penn. State Coll. Miner. Ind. Expt. Sta. Bull. 12 (1933).
[12] N. Johnston, C.M. Beeson, Water permeability of reservoir sands, Trans. Aime 160 (1) (1945) 43−55.

[13] P.H. Monaghan, R.A. Salathiel, B.E. Morgan, et al., Laboratory studies of formation damage in sands containing clays, 216 (12) (1959) 209–215.
[14] Y. Jienian (Ed.), Drilling Fluid Technology, Petroleum University Press, Dongying, 2001.
[15] M. Zhao, T. Xu, et al. (Eds.), Technology for Protecting Oil and Gas Layers, Petroleum Industry Press, Beijing, 1995.
[16] X. Dong, S. Cao, H. Tang, Offshore Drilling Manual, Petroleum Industry Press, 2011.
[17] S. Fan, J. Yan, D. Zhou, Drilling Fluid Completion Fluid and Technology for Protecting Oil and Gas Layers, Petroleum University Press, Dongying, 1996.
[18] T. Xu, M. Zhao, Y. Xiong, Technology for Protecting Oil and Gas Layers, second ed., Petroleum Industry Press, 2003.
[19] Z. Zhang, J. Yan, S. Fan, Low-Density Drilling Fluid Technology, Petroleum University Press, 2004.
[20] G. Jiang, C. Hu, Y. Xiong, et al., Research on the general-purpose "oil film" temporary plugging drilling fluid system, J. China Univ. Pet. (Beijing) (Nat. Sci. Ed.) (4)(2006) 53–57.
[21] K. Chellappah, M.S. Aston, A new outlook on the ideal packing theory for bridging solids;, Proceedings of the SPE International Symposium and Exhibition on Formation Damage Control, Society of Petroleum Engineers, 2012.
[22] N. Hands, K. Kowbel, S. Maikranz, et al., Drill-in fluid reduces formation damage, increases production rates, Oil Gas J. 96 (28) (1998).
[23] S. Yong, Y. Jienian, S. Fuying, et al., Application of a new method of graphical optimization of temporary plugging agent in drilling fluid design, Pet. Drill. Technol. 36 (6) (2008) 48–51.
[24] Z. Fengying, Y. Jienian, L. Zhiyong, et al., Research on particle size optimization of temporary plugging agent during drilling, J. Southwest. Pet. Univ. (Nat. Sci. Ed.) 33 (3) (2011) 130–132.
[25] M. Kaeuffer, Determination de L'Optimum de Remplissage Granulometrique et Quelques Proprietes S'y Rattachant, Congres de I'AFTPV, Rouen, 1973.
[26] Z. Shujie, D. Tongwei, D. Lifang, et al., Application of ideal filling theory in Dagang Oilfield, Drill. Completion Fluids 26 (2) (2009) 123–125.
[27] Z. Fengying, Y. Jienian, Y. Guang, et al., Application of a new method of temporary plugging with ideal filling in Tuhaqiudong low-permeability gas field, Drill. Prod. Technol. 32 (6) (2009) 88–90.
[28] X. Tongtai, C. Yonghao, F. Jinghai, et al., Discussion on the technology of general-purpose shielding temporary plugging to protect oil and gas layers, Drill. Fluid Completion Fluid 20 (2) (2003) 39–41.
[29] J. Guancheng, H. Chengliang, X. Ying, et al., Research on the drilling and completion fluid system of the general-purpose temporary plugging protection oil and gas layer, Drill. Prod. Technol. 28 (5) (2005) 101–104.
[30] H. Chengliang, X. Ying, J. Guancheng, Performance evaluation of a new general-purpose temporary plugging agent GPJ, Oil Gas Field Surf. Eng. 24 (11) (2005) 19–20.
[31] X. Ying, H. Chengliang, J. Guancheng, et al., Development of GPJ series general-purpose temporary plugging agent for drilling fluid, Drill. Fluid Completion Fluid 23 (1) (2006) 1–4.
[32] M. Bao, Study and application on the oil-film method used for reservoir protection drilling and completion fluid systems, J. Dispers. Sci. Technol. 31 (9) (2010) 1273–1277.
[33] J. Guancheng, J. Chaofeng, M. Xianping, et al., Design and application of drilling and completion fluid system for protected reservoir in key blocks of dagang Oilfield, Oil Drill. Prod. Technol. 30 (6) (2008) 58–61.

[34] J. Guancheng, M. Xianping, J. Chaofeng, et al., Application of general-purpose "oil film" temporary plugging agent in oil layer protection technology, Appl. Chem. 24 (6) (2007) 665–669.
[35] L. Yang, Research on the Mechanism of Reservoir Damage in Jilin Oilfield, Daqing Petroleum Institute, 2005.
[36] Z. Jinbo, Y. Jienian, Optimization of particle size distribution of drilling fluid temporary plugging agent, Oilfield Chem. 22 (1) (2005) 1–5.
[37] A. Abrams, Mud design to minimize rock impairment due to particle invasion, J. Pet. Technol. 29 (5) (1977) 586–592.
[38] L. Xiangdong, L. Pingya, Application research of shielding temporary plugging technology in reservoir protection, Drill. Fluid Completion Fluid 9 (2) (1992) 19–27.
[39] L. Pingya, K. Yili, M. Yingfeng, Reservoir protection technology in my country has achieved leapfrog development, Nat. Gas Ind. 26 (1) (2006) 84–87.
[40] Z. Shaohuai, L. Pingya, Reservoir Protection Technology, Petroleum Industry Press, 1993.
[41] L. Kexiang (Ed.), Drilling and Completion Technology for Protecting Oil and Gas Layers, Petroleum Industry Press, Beijing, 1993.
[42] W. Renpu (Ed.), Modern Completion Project, Petroleum Industry Press, Beijing, 1996.
[43] L. Shuaiying (Ed.), Protective Oil Layer Technology in Oilfield Development and Production, Petroleum Industry Press, Beijing, 1996.
[44] L. Pingya, Reservoir Protection Technology, Petroleum Industry Press, Beijing, 1999.
[45] W. Jianjun, L. Xiangdong, Research on shielded temporary plugging of low-permeability core, Drill. Fluid Completion Fluid 9 (3) (1992) 29–34.
[46] X. Houying, Research and Application of Drilling and Completion Fluid System for New-Type Protective Oil and Gas Layers in Dagang Oilfield, China University of Petroleum, Beijing, 2008.
[47] W. Jiang, Y. Jienian, H. Kaiwen, et al., Research and application of ideal filling reservoir protection technology in Tuha Oilfield, Oilfield Chem. 32 (2) (2015) 163–168.
[48] C. Yingchun, Z. Yan, Application of fractal geometry theory in optimal screening temporary plugging agent, J. Univ. Pet. (Nat. Sci. Ed.) (2)(2000) 17–20. 10.
[49] L. Xuegang, Z. Jian, Application research on modified straw fiber of temporary plugging agent in oil and gas layer, Oil Drill. Prod. Technol. 26 (6) (2004) 24–26.
[50] P. Yongqiang, Research on Damage Mechanism and Evaluation Technology of Daqing Fractured Reservoir, Daqing Petroleum Institute, 2009.
[51] W. Zhihong, O. Wei, L. Juquan, et al., Application of multi-stage bridge temporary blocking technology in Hechuan block, J. Chongqing Univ. Sci. Technol.: Nat. Sci. Ed. 14 (5) (2012) 77–79.
[52] J. Guancheng, H. Dongliang, G. Xunzhong, et al., Research and application of a new general-purpose "oil film" temporary plugging drilling and completion fluid system, J. Appl. Basic Eng. Sci. 15 (1) (2007) 74–83.
[53] L. Cuiec, M.T. Bieber, C. Jacquin, et al., Determining the wettability of a reservoir-rock sample, Rev. Inst. Franc. du Petrole 33 (5) (1987) 705–728.
[54] A.J. Staverman, Non-equilibrium thermodynamics of film processes, Trans. Faraday Soc. 48 (1952) 176–185.
[55] E. Van Oort, A. Hale, F. Mody, Manipulation of coupled osmotic flows for stabilisation of shales exposed to water-based drilling fluids, Proceedings of the SPE Annual Technical Conference and Exhibition, Society of Petroleum Engineers, 1995.
[56] H. Daquan, X. Shaoying, T. Zengyan, Application of general-purpose shielding temporary plugging protection oil layer technology in Dagang Oilfield, Drill. Fluid Completion Fluid (2004).
[57] L. Baohe, China Oil and Gas Field Development Journal, Petroleum Industry Press, 2011.

[58] Y. Jinrong, Z. Yuci, Research on shielding temporary plugging technology in Xiazijie Oilfield, Oilfield Chem. 10 (2) (1993) 110–115.
[59] Z. Yuci, Y. Jinrong, Application of shielding temporary plugging technology, Drill. Fluid Completion Fluid (4)(1993) 15–18.
[60] Z. Yuci, Y. Jinrong, L. Dejun, et al., Application of shielding temporary plugging technology in Xinjiang Oilfield, Drill. Fluid Completion Fluid 12 (3) (1995) 74–76.
[61] Q. Jianjun, W. Jianjun, Application of shielding temporary plugging technology in Henan Oilfield, Drill. Fluid Completion Fluid 12 (4) (1995) 56–59.
[62] L. Qi, G. Jianming, Quantitative design of shielding temporary plugging process, Drill. Fluid Completion Fluid 12 (5) (1995) 33–35.
[63] X. Liangcai, G. Wenlin, Tuha Oilfield reservoir protection technology, Drill. Fluid Completion Fluid 13 (5) (1996) 39–41.
[64] Y. Tongyu, Z. Furen, Optimal selection of temporary plugging agent particle size in shielding temporary plugging technology, Fault Block Oil Gas Field 3 (6) (1996) 50–54.
[65] J. Sun, Research on Water-based Drilling Fluid Film Formation Technology, Southwest Petroleum University (2006).
[66] C. Shituan, Y. Nengchuan, Mathematical model and application of shielded temporary plugging technology, Oil Drill. Technol. 25 (1) (1997) 19–20.
[67] Y. Hengshen, M. Bozhong, Optimal calculation method of drilling fluid solid phase particles in shielded temporary plugging technology, Drill. Fluid Completion Fluid 14 (4) (1997) 6–9.
[68] L. Desheng, Z. Zuguo, Application of shielding temporary plugging technology in Huabei Oilfield, Oil Drill. Prod. Technol. 21 (4) (1999) 38–42.
[69] H. Jingrong, Research on the technology of shielding temporary plugging to protect oil and gas layers by modified drilling fluid, Drill. Prod. Technol. 23 (4) (2000) 70–73.
[70] Z. Guoxin, Z. Shanbo, C. Gongsheng, et al., Application of shielding temporary plugging technology in adjustment wells in Shuanghe Oilfield, Pet. Geol. Eng. 15 (1) (2001) 33–35.
[71] L. Xiang, C. Dinglin, Development and performance evaluation of a high-strength composite temporary plugging agent, Chem. Eng. 27 (11) (2013) 46–49.
[72] R. Ewy, R. Stankovich, Pore pressure change due to shale-fluid interactions: measurements under simulated wellbore conditions, Proceedings of the 4th North American Rock Mechanics Symposium, American Rock Mechanics Association, 2000.
[73] J. Zhang, T.M. Al-Bazali, M.E. Chenevert, et al., Factors controlling the film efficiency of shales when interacting with water-based and oil-based muds, SPE Drill. Completion 23 (2) (2008) 150–158.
[74] C.E. Osuji, M.E. Chenevert, M.M. Sharma, Effect of porosity and permeability on the film efficiency of shales, Proceedings of the SPE Annual Technical Conference and Exhibition, Society of Petroleum Engineers, 2008.
[75] H. Zhang, et al., Research and application of drilling fluids to mitigate porosity impairment of deep and low porosity reservoirs, Drill. Fluid Completion Fluid 26 (4) (2009) 4–7.
[76] G. Bo, Research on Damage Mechanism and Protection Technology of Fractured Ultra-Low-Permeability Sandstone Reservoirs, Southwest Petroleum Institute, 2005.
[77] J. Wen, Z. Jun, W. Jianjun, et al., Temporary plugging technology and application of reservoir shielding in the border city area of northern Jiangsu, Nat. Gas Ind. 26 (3) (2006) 81–83.

[78] Y. Jian, K. Yili, L. Lin, et al., Laboratory research on shielding temporary plugging technology for protecting fractured tight sandstone reservoir in central Sichuan, Drill. Fluid Completion Fluid 23 (4) (2006) 28–30. 90–91.
[79] L. Dongtao, Fine Characterization and Quantitative Characterization of the Micro-Pore Structure of Low-Permeability Reservoirs, China University of Petroleum, East China, 2014.
[80] J. Guancheng, W. Shuqi, Application of new shielding temporary plugging technology in Dawanqi area, Oil Drill. Technol. 27 (6) (1999) 21–23.
[81] L. Jun, X. Shaoying, W. Changbing, et al., Application of general-purpose shielding temporary plugging technology in Dagang Oilfield, Drill. Prod. Technol. 27 (5) (2004) 18–21.
[82] H. Jiliang, L. Xuezhang, L. Sihai, Application research of general-purpose temporary plugging technology in sandstone reservoir of Zhunzhong 1 area, Prospecting Eng. (Rock. Soil. Drill. Eng.) 32 (6) (2005) 46–48.
[83] G. Dingxiong, R. Xun, K. Shaohua, et al., Application of general-purpose shielding temporary plugging technology in Yulou area of Liaohe, Foreign Well Logging Technol. (4)(2013) 56–58.
[84] J. Guancheng, Theoretical Basis and Application of Gas Wettability on Rock Surface of Porous Media Reservoirs, China University of Petroleum Press, Dongying, 2015.
[85] J. Guancheng, X. Yang, W. Xi, et al., Wetting reversal agent and reservoir protection agent composition suitable for low-permeability ultra-low-permeability reservoirs and their applications, ZL 201510569878.1.
[86] J. Guancheng, Z. Xianmin, W. Le, H. Yinbo, L. Fan, Y. Lili, et al., Double cationic fluorocarbon surfactant and its preparation method and as double sparse type application of wetting reversal agent and drilling fluid and its application, ZL 201710038133.1.
[87] G. Jiang, Y. Xuan, X. Wang, S. Zhang, Y. An, J. Sun, A preparation method of a wettability reversal agent, US 9, 296, 936.
[88] Q. Yubin, Research on Preparation and Wetting Characteristics of Silicon-Based Bionic Self-Cleaning Surface, Qingdao Technological University, 2012.
[89] S. Na, W. Juntao, J. Lei, Research progress of shell nacre and its bionic materials, Chem. J. Chin. Univ. 32 (10) (2011) 2231–2239.
[90] G. Jiang, Y. Xuan, X. Wu, et al., Method for preparation of bionic polymer for stabilizing wellbore and drilling fluid, US9410068, 2016.
[91] G. Jiang, Y. Xuan, X. Zhang, et al., Bionic drilling fluid and preparation method thereof, US9528041, 2016.

CHAPTER 5

New technology of bionic drilling fluids featured by bubble wrapping and cuttings carrying

As a result of the ongoing development of complex oil and gas reservoirs, wells with complex structures such as horizontal wells and extended reach wells have gradually become the main well types for exploitation. Owing to the special well structure of such complex wells, during the drilling process, gravity causes the cuttings to gradually accumulate in the deviated and horizontal sections, forming a cuttings bed. The presence of the cuttings bed seriously affects the drilling safety and efficiency of such complex wells. Therefore solving the problem of cuttings carrying has become an important issue for drilling complex wells [1–6].

5.1 The principle of water spiders' using superhydrophobicity to achieve the efficient adhesion of bubbles and its bionic application in drilling fluids

Water spiders live completely in water, and rely on their abdomen to efficiently adhere air bubbles so that they can effectively breathe, hunt, and lay eggs in water. This unique characteristic has aroused great interest from researchers, and in-depth research has been conducted. It has been found that water spiders' bubble adsorption ability stems from the superhydrophobicity of the abdomen. Analysis of the microscopic morphology of natural biological surfaces such as water spiders, dragon lice, paddle beetles, and the chemical properties of the surface shows that the structure of this superhydrophobic surface is affected not only by the surface micromorphology, but also by the surface chemical properties. The combination of the two aspects maximizes the hydrophobic properties of the surface. Inspired by this principle, considerable bionic research has been conducted, and many new materials have been developed, such as fluorination on the surface of carbon nanotubes and polyurethane curing on the surface to achieve superhydrophobic properties. These technologies

Fundamentals and Applications of Bionic Drilling Fluids
DOI: https://doi.org/10.1016/B978-0-323-90293-9.00005-8

have extensive applications on metal surfaces, and they also have surface modification effects on paper and cotton [7–15]. However, these studies have focused on the simple modification of solid surfaces and have not yet involved applications in formations. During the drilling process in the petroleum industry the broken cuttings of the drill bit are carried to the surface by the drilling fluid to be removed, keeping the borehole clean. However, in some cases, in drilling complex structural wells such as horizontal wells, multilateral wells, and fishbone wells or when the penetration rate is fast and a large amount of cuttings is generated per unit time, it is difficult for the cuttings to be carried and suspended and difficult to keep the wellbore clean. Cuttings beds form easily, bringing risks to downhole safety. Drilling operations often combine methods such as well washing, hole reaming, back reaming, and short-range round trip to prevent cuttings from being deposited at the bottom of the borehole, but the mechanical method has the disadvantages of long operation time and high cost [16–32]. In recent years, low-density, low-buoyancy, and high-viscosity foam drilling fluids have been applied to hole cleaning, but they have little effect on hole cleaning and lead to excessive cuttings deposition [33]. Yu et al. first proposed the use of surfactants to conduct cuttings flotation and clean the annulus space of wellbore, and the effect is good, but they did not explain the type of surfactant used and its mechanism [34–41]. Dong et al. [42] used GOB cuttings-carrying technology in the process of drilling the Jing 52-H1Z multilateral well. Some bubbles were generated from mineral oil and surfactants and oil beads, which adhered to the surface of the cuttings and dragged the cuttings upward, thereby achieving the purpose of suspending and carrying cuttings, but the effect was not evaluated [43–48]. Adjusting the rheological parameters of drilling fluids and increasing the displacement of mud pumps are other common methods used to achieve hole cleaning, but the results are often unsatisfactory. In response to the shortcomings of traditional technologies, the team led by Professor Guancheng Jiang of China University of Petroleum (Beijing) developed a bubble-wrapping and cuttings-carrying agent FGC-1 based on water spiders' adhering to bubbles. This enabled development of the new technology of bionic drilling fluids featured by cuttings-carrying hole cleaning. The effect of cuttings carrying was tested by designing a wellbore simulation device for extended reach wells. The results show that this new technology effectively solved the problem of cuttings carrying in complex wells such as horizontal wells and extended reach wells [49].

5.2 Development and evaluation of FGC-1

5.2.1 Development of FGC-1

In a 500-mL four-necked flask equipped with an electric stirrer, a condenser tube, a nitrogen introduction tube, and a thermometer, a series of monomers such as fluorine-containing monomers were slowly added to the anionic surfactant emulsification system, with nitrogen aerated for protection. The temperature of water bath was raised to 65°C, and the APS aqueous solution was added to initiate the copolymerization. After the self-temperature increase process was completed, the product was stirred at a moderate speed and a constant temperature of 80°C for 4 h. After the reaction was completed, the temperature was naturally lowered to below 60°C, and the pH value was adjusted to neutral with Na_2CO_3. The product was further cooled to below 45°C. The stirring stopped, and the emulsion was filtered and poured out to obtain the FGC-1.

5.2.2 Evaluation of the effect of FGC-1

5.2.2.1 Effect of FGC-1 on core surface wettability

The core was immersed in aqueous solutions of different concentrations prepared with FGC-1 for 4 h. The core was naturally dried at room temperature for 30 min to remove the excess solvent on the surface of the core. A JC2000D3 contact angle measuring instrument was used to determine the contact angle of distilled water and n-hexadecane on the core

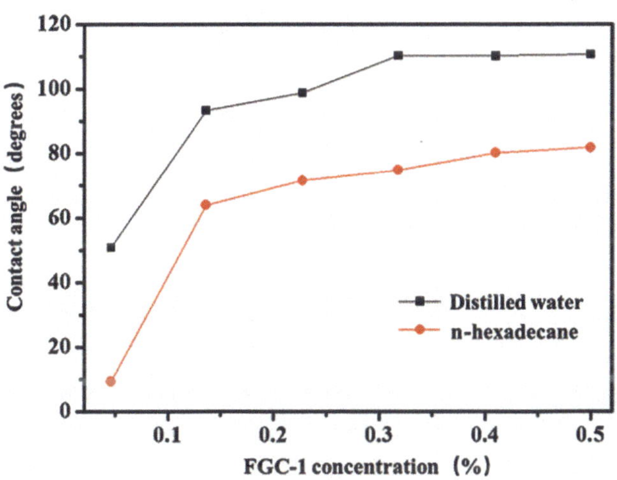

Figure 5.1 Contact angles of water and n-hexadecane on the core surface treated by the FGC-1 solution at different concentrations.

surface, with the results shown in Fig. 5.1. After the core was treated with FGC-1, the surface wettability obviously reversed. With the increase of FGC-1 concentration the surface of the core gradually changed from lipophilic and hydrophilic to oil- and water-repellent, and the degree of transformation became more and more obvious. When the concentration of FGC-1 was 0.5%, the contact angle of distilled water reached 110.5 degrees, and that of n-hexadecane reached 79.0 degrees. It can be seen that FGC-1 can change the wettability of the core surface, making it oil- and water-repellent and get contact with gas in priority.

5.2.2.2 Effect of FGC-1 on bubble adhesion performance

To further confirm that FGC-1 can enable the surface of the cuttings to have good bubble adhesion performance, two commercial surfactants were selected and compared with FGC-1. The experimental steps were as follows: At room temperature, three beakers with a magnetic rotor were selected, followed by the addition of surfactant 1, surfactant 2, FGC-1 prepared in advance at a certain concentration, and a proper amount of foaming agent solution. Then 40 g of cuttings were added to each beaker, air bubbles were generated by aeration, and the magnetic stirrer was turned on. After 5 min, the aeration of air bubbles was stopped, and the magnetic stirrer was also stopped. The products were left standing for a period of time. The experimental results are shown in Fig. 5.2. It can be seen that at first, a small amount of bubbles adhered to the surface of the cuttings in the beaker with ordinary surfactants, and there were almost no air bubbles after a period of time. By contrast, a large number of bubbles adhered to the surface of the cuttings in the FGC-1 solution, as shown by the red circle in Fig. 5.2. The rotation of the magnetic rotor drove the cuttings to move in the beaker. With the continuous aeration of bubbles, the moving cuttings collided with the bubbles, adhered together, and rose to the surface of the water with the bubbles. In addition, the foaming agent enhanced the strength of the bubbles such that they are not easily broken.

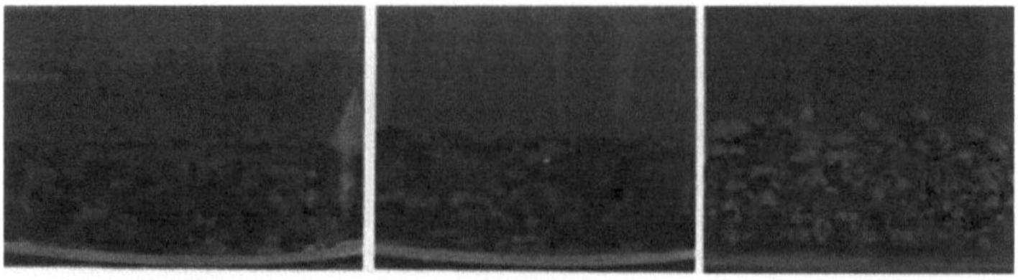

Figure 5.2 Comparison of the effect of bubbles adhering to cuttings.

5.2.2.3 Effect of FGC-1 on carrying the cuttings with different particle sizes

To investigate the effect of FGC-1 on carrying cuttings with different particle sizes, several samples of FGC-1 solution with a concentration of 0.4% were prepared, and an appropriate amount of foaming agent was added to make the cores into cuttings with particle sizes of 2.8–3.2, 2.0–2.8, 0.9–2.0, and 0.45–0.90 mm. The experimental results are shown in Fig. 5.3. Bubbles adhered to the surface of the cuttings, and they floated up to the surface of the solution under the effect of buoyancy. The cuttings with a particle size of 0.9–2.0 mm were easily adhered by the bubbles and were carried to the liquid surface in large quantities and aggregated into clusters. A small number of cuttings with particle size larger than 2.00 mm or less than 0.90 mm was carried by bubbles, most of which were deposited on the bottom of the beaker, and bubbles were also adhered on the surface.

5.2.2.4 The influence of pH value on the cuttings-carrying effect after FGC-1 treatment

The cuttings with a particle size of less than 0.85 mm were selected as the research object for preparing the aqueous solution of different pH values. The cuttings after FGC-1 treatment were placed in the solution, the

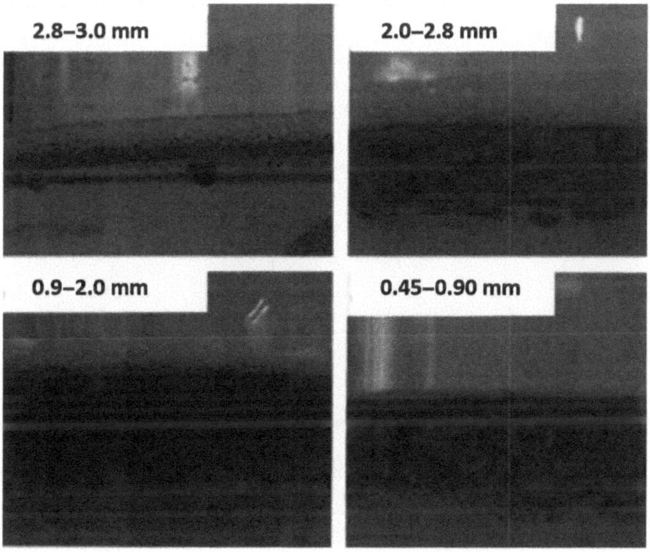

Figure 5.3 Effect of FGC-1 on cuttings carrying with different particle sizes.

suspension of the cuttings was observed, and the mass of the suspended cuttings was weighed to calculate the percentage of suspended cuttings. The influence of pH on the cuttings-carrying effect was investigated, and the results are shown in Fig. 5.4. As the pH of the solution changed, the cuttings suspension performance changed significantly after FGC-1 treatment. As the pH increased, the percentage of suspended cuttings increased from 30.8% at pH = 1 to 96.2% at pH = 3. The pH value continued to increase, and the percentage of suspended cuttings decreased slowly first. When pH > 11, the percentage of suspended cuttings decreased rapidly from 86.6% to 51.4%.

It can be seen from Fig. 5.5 that the absolute value of the zeta potential on the surface of the cuttings first decreased and then increased as the pH increased. When the pH was around 3, the absolute value of the zeta potential was 0, and the surface of the cuttings was not charged, with the least stress in the entire solution system, and the cuttings were most likely to suspend on the liquid surface. As the pH value continued to increase, the absolute value of the zeta potential on the cutting surface continued to increase, and the unstable cutting force weakened the cuttings suspension performance. When the pH value was greater than 11, the hydration film on the surface of the cuttings became thicker, and the increase of the hydration film increased the instability of the cuttings, thereby greatly enhancing the sedimentation ability of the cuttings, which was similar to the results in Fig. 5.5.

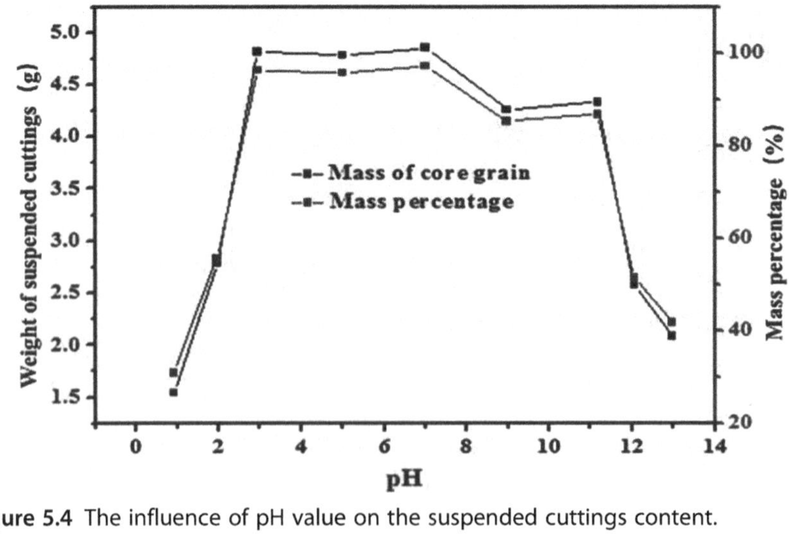

Figure 5.4 The influence of pH value on the suspended cuttings content.

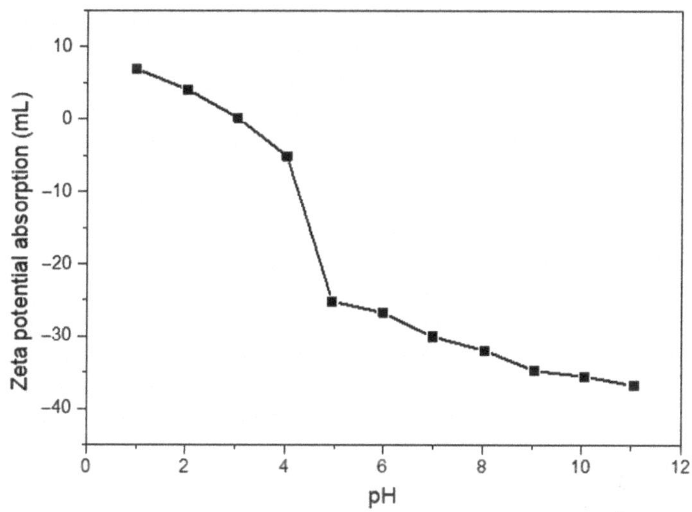

Figure 5.5 The influence of zeta potential on the surface of cuttings with the solution pH value.

5.2.2.5 Effect of cation concentration on the cuttings-carrying effect after FGC-1 treatment

Cuttings with a particle size of less than 0.85 mm were selected as the research object. The aqueous solutions of different pH values were prepared. The cuttings after FGC-1 treatment were placed in the solution, the suspension of the cuttings was observed, and the mass of the suspended cuttings was weighed to calculate the percentage of suspended cuttings. The effect of cuttings carrying on the sodium ion concentration was investigated. The results are shown in Fig. 5.6. With the increase of sodium ion concentration, the percentage of suspended cuttings after FGC-1 treatment showed a trend of increasing first and then decreasing. When the sodium ion concentration was 0.8 mol/L, the percentage of suspended cuttings reached its maximum (98%). As the sodium ion concentration continued to increase to 1.3 mol/L, the percentage of suspended cuttings decreased to 87.2%.

As can be seen from Fig. 5.7, the absolute value of the zeta potential on the surface of the cuttings gradually decreased at first and then increased gradually with the increase of the sodium ion concentration. When the sodium ion concentration was 0.8 mol/L, the zeta potential on the surface of the cuttings was 0. At this time, the surface of the cuttings reached the isoelectric point, and the cuttings were least affected by the force. This was consistent with the results in Fig. 5.6 when the sodium ion concentration reached 0.8 mol/L.

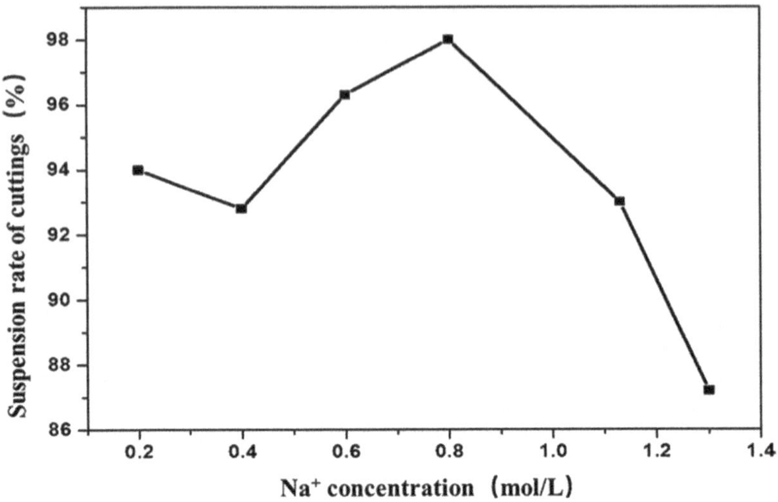

Figure 5.6 The influence of suspension rate of cuttings with Na^+ concentration.

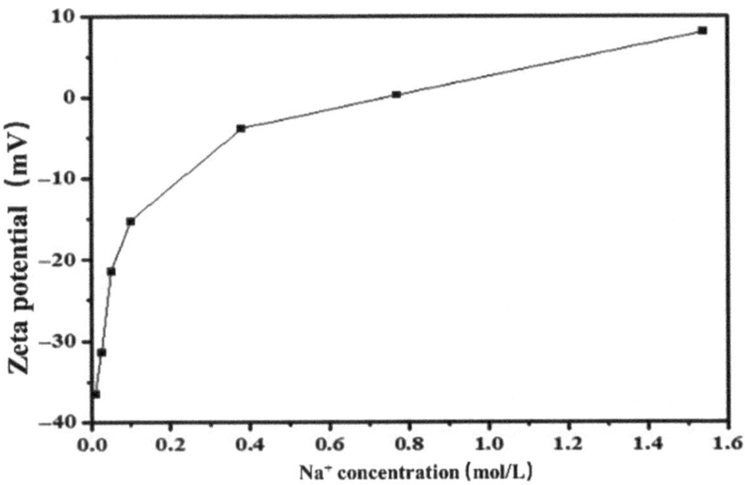

Figure 5.7 Influence of zeta potential on the surface of cuttings with Na^+ concentration.

5.2.3 Analysis of the mechanism of FGC-1

The experimental results show that FGC-1 can be fully adsorbed on the core surface, changing the core surface wettability, making it oil- and water-repellent and preferentially contact with the gas, that is, achieving the gas wettability of the cuttings. As a result, the relative density of cuttings is reduced, which further improves the suspendability and effectively improves the cuttings-carrying capacity. The microscopic cuttings-carrying mechanism of FGC-1 is shown in Fig. 5.8.

FGC-1 is a surface-active substance composed of polar/nonpolar molecules. Its polar end can be selectively adsorbed on the surface of hydrophilic cuttings, and the nonpolar end faces water. In this way, the surface of the hydrophilic cuttings is converted into a hydrophobic substance, which sticks to the bubbles. In the simulation device, after the air passes through the metal screen, small bubbles are formed in the drilling fluid. The small bubbles come into contact with the cuttings particles above the screen and collide to form a cuttings-bubble aggregate. At the same time, they are affected by buoyancy, which relatively reduces the weight of the cuttings. If the weight of the cuttings is less than the buoy ancy effect, the cuttings will be brought up by bubbles under the effect of buoyancy. In addition, the bubbles are adsorbed on the cuttings surface, whose surface area is therefore increased. Under the action of fluid flow, the drag force of the cuttings is enhanced, so in the actual drilling process, the drill pipe is stirring in the wellbore, and the cuttings deposited at the bottom of the well will be driven due to the adhesion of bubbles on the surface and circulated out of the wellbore with the drilling fluid flow.

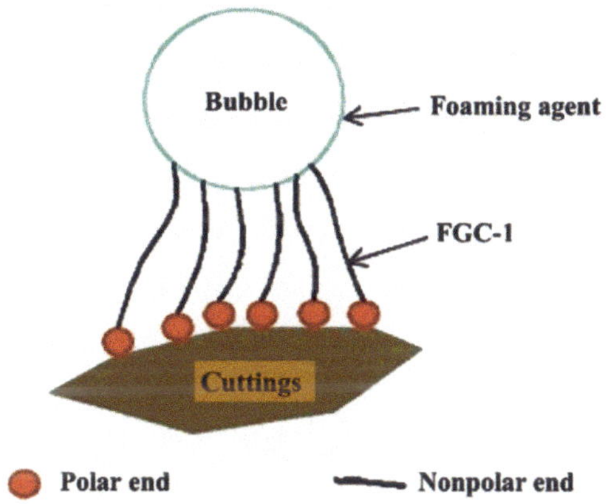

Figure 5.8 Microscopic mechanism of FGC-1.

5.3 Establishment of the new technology of bubble-wrapped bionic drilling fluids

5.3.1 Influence of FGC-1 on the rheological properties of drilling fluids

Two sets of drilling fluid systems were selected to investigate the effect of FGC-1 on their rheological properties. The drilling fluid formulas are as follows:

1#(Brine drilling fluid system):0.5% bentonite + 0.3% PAC − LV + 9% GJL − 1 + 0.5% FA − 367 + 3% SMP − II + 3% SPNH + 2% FT − 1A + 3% polymerization alcohol (lubricant) + 7% KCl + 22% NaCl + 0.65% NaOH + 0.1% XC, stirring for 2 h.

2#(Solid-free film-type drilling fluid system):2% KH − 931 + 0.1% silicone viscosity reducer IV + 2% low softening point asphalt + 2% solid lubricant + 0.5% PAC − LV + 0.1% FA − 367 + 1% filming agent (made in the laboratory) + 0.17% XC, adjust pH to 10, and stir for 2 h.

1. The effect of FGC-1 on the rheological properties of drilling fluid at room temperature

   The effect of different concentrations of FGC-1 on the rheological properties of drilling fluid at room temperature is shown in Table 5.1. After FGC-1 was added to brine drilling fluid and solid-free filming drilling fluid, both the viscosity and the YP increased with the increase in FGC-1 concentration. When the concentration of FGC-1 was 0.4%–0.5%, the drilling fluid had good rheological properties.

Table 5.1 Effect of different concentrations of FGC-1 on the rheological properties of drilling fluids.

Drilling fluid	FGC-1 (%)	AV (mPa · s)	PV (mPa · s)	YP (Pa)	YP/PV (Pa/mPa · s)
1	0	22.60	16.2	6.40	0.40
	0.1	24.30	14.4	9.90	0.69
	0.2	25.10	15.2	9.90	0.69
	0.3	25.70	13.8	11.90	0.86
	0.4	26.00	15.6	10.40	0.67
	0.5	28.05	17.2	10.85	0.63
2	0	27.60	19.0	8.60	0.45
	0.1	29.80	21.6	8.20	0.38
	0.2	31.25	22.2	9.05	0.41
	0.3	34.10	23.2	10.90	0.47
	0.4	37.20	28.2	9.00	0.32
	0.5	40.00	26.4	13.60	0.52

2. The effect of FGC-1 on the rheological properties of drilling fluid after hot rolling

The rheological measurement results of the two systems after hot rolling at 100°C for 16 h are shown in Table 5.2. Under different dosages of FGC-1, the viscosity and YP of brine drilling fluid decreased significantly, while the viscosity and YP of the solid-free filming drilling fluid changed little in comparison with those before hot rolling.

5.3.2 Evaluation of the cuttings carrying effect of FGC-1 in the simulation device

The cuttings-carrying effect of a brine drilling fluid system containing different concentrations of FGC-1 was measured, as shown in Fig. 5.9. It can be seen that when FGC-1 was not added, the drilling fluid was not ideal for the carrying of cuttings. When the particle size of the cuttings was fixed, as the concentration of FGC-1 increased, the amount of cuttings carried by the bubbles to the liquid surface gradually increased. When the concentration of FGC-1 was 0.4% and the particle size of the cuttings was between 0.90 and 2.00 mm, the carried cuttings accounted for 77.2% of the total mass. When the concentration of FGC-1 continued to increase to 0.5%, there was no obvious increasing trend in the cuttings-carrying effect. It can be seen that when the FGC-1 concentration was controlled to 0.4%, the carrying effect on

Table 5.2 Effect of FGC-1 on the rheological properties of drilling fluids after hot rolling (100°C, 16 h).

Drilling fluid	FGC-1 (%)	PV (mPa · s)	YP/Pa	YP/PV (Pa/mPa · s)
1	0	7.6	4.4	0.58
	0.1	8.1	5.5	0.68
	0.2	7.8	6.2	0.79
	0.3	8.2	6.6	0.80
	0.4	9.7	5.9	0.61
	0.5	10.8	6.3	0.58
2	0	17.4	9.3	0.53
	0.1	16.2	9.3	0.57
	0.2	17.4	10.4	0.60
	0.3	19.2	12.5	0.65
	0.4	20.2	13.9	0.69
	0.5	20.3	14.95	0.74

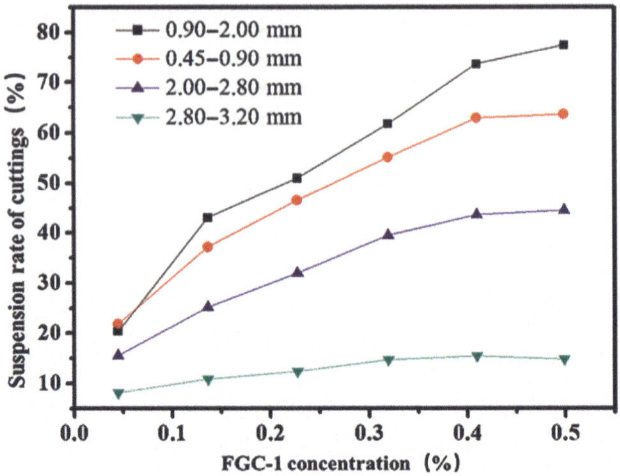

Figure 5.9 FGC-1's cuttings-carrying effect with different particle sizes.

the cuttings with a particle size of 0.90–2.00 mm was the best, and it was cost-effective and efficient.

In summary, based on the model that water spiders rely on superhydrophobic fluff to efficiently adhere bubbles under water to provide buoyancy, combined with the developed theory of rock gas wettability, FGC-1 featured by bubble wrapping and cuttings carrying was developed, and a new technology of bionic drilling fluids featured by bubble wrapping and cuttings carrying was established. Through a series of effect evaluation and mechanism analysis, the following conclusions are obtained:

1. FGC-1 can be fully adsorbed on the core surface, changing the wettability of the core surface, making it oil- and water-repellent and preferentially contacting the gas, that is, achieving the gas wettability of the cuttings surface, reducing the relative density of the cuttings, and improving the cuttings-carrying capacity of the drilling fluids.
2. Compared with ordinary surfactants, FGC-1 can make the bubbles adhere to the surface of the cuttings permanently and uniformly, showing superior bubble adsorption performance.
3. The optimal concentration of FGC-1 is 0.4%, and the optimal particle size of cuttings is 0.90–2.00 mm, which can carry the cuttings accounting for 77.2% of the total mass to the solution surface.
4. After FGC-1 is added to the drilling fluid, the rheological properties of the drilling fluid before and after hot rolling meet the requirements for on-site construction,

5. The cuttings are well carried and transported in the wellbore simulation device, and bubbles adhere to the surface of the cuttings. A large amount of the cuttings are circulated to the water outlet as the drill string rotates and the drilling fluid flows.

References

[1] W. Feng, The advantages of drilling technology for complex oil and gas reservoirs are highlighted, PetroChina (9)(2005) 59.
[2] L. Shuangliang, Research on Wellbore Purification of Extended Reach and High Slope Wells, Yangtze University, 2012.
[3] D. Yan, Research on Optimization of Wellbore Purification Technology for Marine Extended Reach Drilling, Xi'an Shiyou University, 2017.
[4] M.E. Ozbayoglu, M. Sorgun, A. Saasen, Hole cleaning performance of light-weight drilling fluids during horizontal underbalanced drilling, J. Can. Pet. Technol. 49 (4) (2010).
[5] T.I. Larsen, A.A. Pilehvari, J.J. Azar, Development of a new cuttings-transport model for high-angle wellbores including horizontal wells, SPE Drill. Completion 12 (1997) 129–136.
[6] J. Li, S. Walker, Sensitivity analysis of hole cleaning parameters in directional wells, SPE J. 6 (4) (2001) 356–363.
[7] Y. Long, Z. Pu, X. Huang, et al., Effect of CuPc@MWCNTs on rheological, thermal, mechanical and dielectric properties of polyarylene ether nitriles (PEN) terminated with phthalonitriles, J. Polym. Res. 21 (9) (2014) 525.
[8] L. Mingzhong, W. Weiyang, H. Yanfeng, et al., Study on the sand-carrying law of vertical wellbore, J. China Univ. Pet. Ed. Nat. Sci. 24 (2) (2000) 33–35.
[9] L. Gang, W. Xia, X. Xiqing, et al., Application of two-phase flow theory in sand-carrying in gas wells, Fault Block Oil Gas Fields 15 (6) (2008) 65–67.
[10] W. Changjiang, Analysis of the factors affecting the concentration of cuttings in foam drilling annulus, Nat. Gas Ind. 30 (3) (2010) 83–86.
[11] J. Heping, The effect of foam fluid on cuttings, Geol. Explor. 12 (1990) 56.
[12] D. Shaoqiang, H. Ming, X. Li, et al., Research on sand-carrying capacity of sand-producing gas wells, Spec. Oil Gas Reserv. 15 (3) (2008) 87–89.
[13] Z. Shuping, W. Xiaorong, F. Lianlian, et al., Research and application of sand-carrying theory in gas wells, Fault Block Oil Gas Field 14 (1) (2007) 50–52.
[14] W. Zhimin, H. Fengtao, J. Qingsheng, et al., Preliminary exploration on the feasibility of heavy oil carrying sand production, Oil Drill. Prod. Technol. 20 (2) (1998) 104–105.
[15] H. Lianyin, S. Xiutong, H. Guoyuan, Production technology of sand-carrying oil wells carrying sand, Oil Drill. Prod. Technol. 21 (1) (1999) 098–99101.
[16] M.S. Bizanti, S.F. Alkafeef, A simplified hole cleaning solution to deviated and horizontal wells, 2003. Society of Petroleum Engineers Middle East Oil Show, SPE 81412 (n.d.).
[17] S.A. Mirhaj, R. Shadizadeh, M. Fazaeli-Zadeh, Cuttings removal simulation for deviated and horizontal wellbores, in: SPE Middle East Oil and Gas Show and Conference, 2007.
[18] C. Hyun, S. Shah, S. Osisanya, A three-layer modeling for cuttings transport with coiled tubing horizontal drilling, in: SPE Technical Conference and Exhibition, 2000.
[19] P. Doron, M. Simkhis, D. Barnea, Flow of solid-liquid mixtures in inclined pipes, Int. J. Multiph. Flow. 23 (2) (1997) 313–323.

[20] R. Dave, W. Thomas, A.C. Chong, Steerable hole enlargement technology in complex 3D directional wells, 2003, Society of Petroleum Engineers SPE Asia Pacific Oil and Gas Conference and Exhibition, SPE 80476 (n.d.).
[21] T. Nazari, G. Hareland, J. Azar. Review of cuttings transport in directional well drilling: systematic approach, in: SPE Western Regional Meeting, 2010.
[22] M. Linhu, Experimental study on cuttings carrying in directional and horizontal wells, Oil Drill. Technol. 11 (2) (1989) 58–61.
[23] W. Zhiming, Z. Yujia, G. Qingchun, The application of extended reach well borehole cleaning monitoring technology in Dagang Oilfield, Oil Drill. Prod. Technol. 34 (2) (2012) 17–19.
[24] S. Xuncheng, W. Zhenfei, X. Xiaolong, et al., Research on hazards and treatment measures of cuttings beds in extended reach wells, Pet. Sci. Technol. Forum 31 (2) (2012) 40–42.
[25] X. Tianyong, P. Daozhong, H. Lixin, et al., Mechanism analysis and countermeasures of common problems in extended reach horizontal wells, Drill. Prod. Technol. 24 (5) (2001) 22–25.
[26] W. Haige, L. Xisheng, Experimental study of mud performance and influence of annulus cuttings migration, Drill. Fluid Completion Fluid 11 (2) (1994) 24–30.
[27] Y. Jiafeng, Numerical simulation of annulus pressure loss and cuttings transport law in horizontal section, Pet. Mine Machin. (3)(2014) 57–59.
[28] C. Bin, T. Haixiong, L. Junfeng, et al., Introduction of the first domestic application of the Western Hercules deepwater drilling platform, Ocean Eng. 28 (1) (2010) 125–131.
[29] L. Ming, J. Yafeng, W. Zaiqiang, et al., Numerical simulation of rotation factors of extended reach horizontal wellbore purification, West. China Explor. Eng. 22 (5) (2010) 51–54.
[30] L. Qi, L. Qing, W. Liupeng, et al., Inclined wellbore purification model and its application research, J. Xi'an Shiyou Univ. (Nat. Sci. Ed.) 25 (2) (2010) 39–43.
[31] C. Jun, L. Xisheng, et al., Experimental study of annulus-carrying rocks in horizontal wells, J. Univ. Pet. Nat. Sci. Ed. 8 (4) (1992) 26–30.
[32] W. Haige, L. Xisheng, Experimental study on drilling fluid carried by cuttings in horizontal well section, Acta Pet. Sin. 16 (4) (1995) 125–132.
[33] F. Xiaoduan, Research on Insects of Chinese Gyrinidae in the Collection of Sun Yat-sen University, Sun Yat-sen University, 2005.
[34] A.L. Martins, A.M.F. Lourenco, C.H.M. de Sa, et al., Foam property requirements for proper hole cleaning while drilling horizontal wells in underbalanced conditions, SPE Drill. Completion (2001) 195–200.
[35] L. Jianzeng, Experiment and Simulation Study on Sand-Carrying Capacity of Heavy Oil in Horizontal Wells, China University of Petroleum (Beijing), 2012.
[36] W. Xugang, Study on the Mechanism of Limited Sand-Carrying Cold Recovery in Sudan 6 Heavy Oil Reservoir, Graduate School of Chinese Academy of Sciences, 2010.
[37] Z. Haiying, Screw pump heavy oil carrying sand cold recovery technology, Sci. Technol. Innov. Appl. (2)(2012) 15.
[38] Z. Weihong, Underbalanced foam drilling technology, Nat. Gas Ind. 19 (5) (1999) 51–54.
[39] P. Xiaolin, L. Gang, L. Xingshu, et al., Drilling fluid isolation membrane theory and film-forming drilling fluid research, Drill. Fluid Completion Fluid 22 (6) (2005) 1–4.
[40] Z. Zhenhua, D. Wei, D. Chunxu, et al., Jing 52-H1Z branch horizontal well drilling fluid technology, in: 2010 National Symposium on New Technology Exploration and Development and Drilling and Production Technology of Complex Geological Oil and Gas Fields, 2010.

[41] M. Yu, D. Melcher, N. Takach, et al., A new approach to improve cuttings transport in horizontal and inclined wells, SPE 90529, 2004.
[42] D. Wei, Y. Lie, J. Wenbo, et al., Drilling fluid technology of Jing 52-H1Z branch horizontal well, Drill. Fluid Completion Fluid 26 (3) (2009) 1–4.
[43] W. Zaiming, Research on the Characteristics of Supercritical Carbon Dioxide Drilling Fluid, China University of Petroleum, 2008.
[44] Z. Huansheng, Z. Bo, Feasibility study of supercritical CO_2 fluid rock-carrying, China Pet. Chem. Stand. Qual. 31 (12) (2011) 29.
[45] W. Ruihe, H. Hongjun, N. Hongjian, Research on Rock-Carrying Mechanism in Supercritical Carbon Dioxide Wellbore, National Hydrodynamics Seminar, 2014.
[46] S. Zhonghou, W. Haizhu, L. Gensheng, Numerical simulation of rock-carrying capacity in horizontal section of supercritical CO_2 drilling, Pet. Explor. Dev. 38 (2) (2011) 233–236.
[47] L. Liangchuan, W. Zaiming, Q. Zhengsong, et al., Rock-carrying characteristics experiment of supercritical carbon dioxide drilling fluid, Acta Pet. Sin. 32 (2) (2011) 355–359.
[48] H. Hongjun, W. Ruihe, N. Hongjian, et al., Research on the rock-carrying rule of supercritical carbon dioxide in horizontal well drilling, Pet. Drill. Technol. (2)(2014) 12–17.
[49] J. Guancheng, L. Ling, X. Weixing, et al., Development and effect evaluation of bubble cuttings carrying agent for extended reach wells, Drill. Fluid Completion Fluid 30 (1) (2013) 5–8.

CHAPTER 6

New technology and application of high-efficacy bionic water-based drilling fluids

As the development of oil and gas fields enters its middle and late stages, the difficulty of drilling operations and the cost of well development are increasing dramatically. Typical highly complex wells include ultradeep wells, high-temperature wells, high-pressure wells, extended reach wells, and deep-water wells. The drilling of these wells often encounters complex situations such as resistance to round trip, pipe sticking, a low mechanical penetration rate, borehole instability, lost circulation. and formation damage. The traditional methods for solving these problems is to use oil-based or synthetic-based drilling fluids. However, harsh environmental constraints and the challenges of deep-water drilling have promoted the development of water-based drilling fluids. To meet the goal of safe, efficient, low-cost, and environmentally friendly drilling, extensive research has been conducted since the 1990s on the inhibitory performance, lubricity, cuttings accumulation, cuttings encapsulating capacity, and penetration rate of water-based drilling fluids. A series of high-performance water-based drilling fluid systems have been developed [1−12]. Among them, typical high-performance water-based drilling fluid systems in China include partially hydrolyzed polyacrylamide potassium salt drilling fluid, calcium alcohol complex water-based drilling fluid, silicate drilling fluid, calcium chloride/polymer drilling fluid, and cation/zwitterion drilling fluid [13,14]. Typical international high-performance water-based drilling fluid systems include the ULTRADRIL and HydraGlyde systems of MI-SWACO and the PureBore system of Innospec. High-performance water-based drilling fluids have been widely used in China, Brazil, Argentina, Colombia, the United States, Australia, Saudi Arabia, and other countries [15] and have achieved good results and economic benefits. However, these high-performance water-based drilling fluid technologies have certain drawbacks. For example, the partially hydrolyzed polyacrylamide potassium salt drilling fluid often appears as a mud-drilled bit and cannot be used in high-rate drilling

Fundamentals and Applications of Bionic Drilling Fluids
DOI: https://doi.org/10.1016/B978-0-323-90293-9.00006-X

where there is high water content and sensitive shale; cationic polymer drilling fluids have a high cost and poor compatibility with anionic drilling fluid agents; and it is difficult to adjust the rheological properties of silicate drilling fluid systems [16]. Overall, there is still a gap between these high-performance water-based drilling fluid technologies and oil-based drilling fluids.

In response to these problems, the team led by Professor Guancheng Jiang of China University of Petroleum (Beijing) has developed a series of new bionic drilling fluid technologies based on the development of bionic drilling fluid materials, the formation conditions encountered, and the characteristics of the formations. These include bionic drilling fluids with different temperatures and densities, with or without salt, and with or without clay. The technical problems encountered in the drilling process can be solved by a synergistic technical approach that combines internal and external causes of solidification, inhibition, plugging, lubrication, amphiphobic treatment, and film formation [17]. Solidification refers to enhancing the cohesion and cementing force between the rock particles of the borehole, improving the strength of the formation rock, and emphasizing the use of internal causes to improve the borehole stability. Inhibition refers to not only suppressing the osmotic hydration of clay particles, but also, more important, suppressing the occurrence of surface hydration. Plugging refers to the sealing of pores of micro-nano and above sizes. Lubrication refers to not only improving the lubricity by changing the characteristics and flow patterns of drilling fluids through the principle of fluid lubrication, but also the combination of active components of lubricant with the free ions in the borehole to produce a smooth film between the surface of the drilling tool and the borehole wall, which effectively controls the inherent eddy currents in the flow interface, effectively reduces the frictional resistance in complex drilling constructions such as the deviated sections of extended reach wells, improves the mechanical penetration rate, extends the life of the drilling tools, reduces operating costs, and improves the overall benefits of drilling, that is, synergistically reducing the frictional resistance and torque from flow and bonded lubrication and increasing the surface energy transmission and the penetration rate [18–24], amphiphobic treatment means that the wettability of the rock surface is reversed from hydrophilic and oleophilic to hydrophobic and oleophobic by an amphiphobic agent, which reverses the capillary suction to capillary resistance, thereby preventing oil and/or water from entering the inside of the capillary tube, avoiding the occurrence of capillary phenomena, permeation hydration, and surface hydration and protecting the formations. Film formation refers to the rapid formation of a film on the surface of the rock wall to prevent the liquid and solid phases of the drilling fluid from entering the formation under the effect of a positive pressure differential to achieve the purpose of protecting the formation [25–29]. These synergistic effects enable the technical performance of bionic water-based drilling fluids to approach or reach that of oil-based drilling fluids.

6.1 New technology of high-efficacy bionic water-based drilling fluids

In the first decades of the 21st century, high-performance water-based drilling fluids that can improve inhibition and lubricity have been regarded as one of the research hotspots, and a certain amount of progress has been made. Professor Guancheng Jiang of China University of Petroleum (Beijing) led a team that has introduced bionics into the field of drilling fluids. The bionic water-based drilling fluids that they have developed not only greatly improved the inhibition and lubricity, but also achieved a transformation from high performance to high efficacy for the first time. High performance is essentially different from high efficacy. High performance means that the drilling fluid system itself has good performance and reaches a high level; high efficacy mainly refers to improvement in the quality and benefit of drilled wells that use this drilling fluid, which achieves the goal of safe, efficient, low-cost, and environmentally friendly drilling. Both the drilling fluid performance and the effect of well completion are very good.

6.1.1 Performance comparison of typical bionic water-based and oil-based drilling fluids

Taking the bionic wellbore strengthener, bionic inhibitor, bionic micro-nano plugging agent, bionic bonded lubricant, and amphiphobic agent as the core agents, a high-performance, low-cost, and environmentally friendly bionic drilling fluid system was formed, and its performance was compared with that of typical oil-based drilling fluids.

The basic formula of typical bionic water-based drilling fluids is as follows: 2%–3% bionic wellbore strengthener + 1%–2% bionic inhibitor + 2%–3% bionic nano plugging agent + 1%–3% bionic bonded lubricant + 0.2%–1% amphiphobic agent + supporting agent (or drilling fluid system in the upper well section). If the drilling encounters a leaky formation, a bionic plugging agent for in-drilling condition is added to the drilling fluid. During the drilling process, if serious leakage occurs, a bionic plugging agent to stop the condition is added to the drilling fluid.

The formula of typical oil-based drilling fluids is as follows: 80% 3# white oil + 3% auxiliary emulsifier + 1% main emulsifier + 4% wetting agent + 20% calcium chloride solution + 1% organic soil + 0.5% YP increaser + 4% ultrafine calcium + 4% ultrafine calcium 2000 mesh + 2.67% plugging and filtration control agent + 0.5% viscosity increasing and filtration control agent + 5% lime + barite.

1. Comparison of basic performance

   A comparison of bionic water-based drilling fluids with the basic performance of typical oil-based drilling fluid showed that the rheological properties and filtration loss properties of bionic water-based drilling fluids were comparable, and even the medium-pressure filtration loss and YP/PV ratio were slightly better than those of oil-based drilling fluid, as shown in Table 6.1.

2. Comparison of inhibitory performance

   The comparison of bentonite's expansion height experiments shows that the bionic water-based drilling fluid system was close to the oil-based drilling fluid in terms of inhibiting the hydration dispersion and expansion performance of shales (Fig. 6.1). From the results of cuttings recovery, it can be seen that the bionic water-based drilling fluid increased the recovery from 3.69% to 97.87%, which was close to 99.27% of the oil-based drilling fluid, as shown in Fig. 6.2 and Table 6.2.

Table 6.1 Basic performance comparison of bionic water-based and oil-based drilling fluids.

Formula	(g/cm^3)	AV (mPa · s)	PV (mPa · s)	YP (Pa)	YP/PV	GEL 10′/10″ (Pa)	FL_{HTHP} (mL)	pH
Bionic water base	2.41	103	82	21	0.26	5.5/25	2.4	9
Oil base	2.41	129.5	105	24.5	0.23	4.5/18	2.4	9

Note: The samples were tested after hot rolling at 130°C for 16 h, and the FL_{HTHP} test temperature was 130°C. Abbreviations: *HTHP*, high temperature and high pressure.

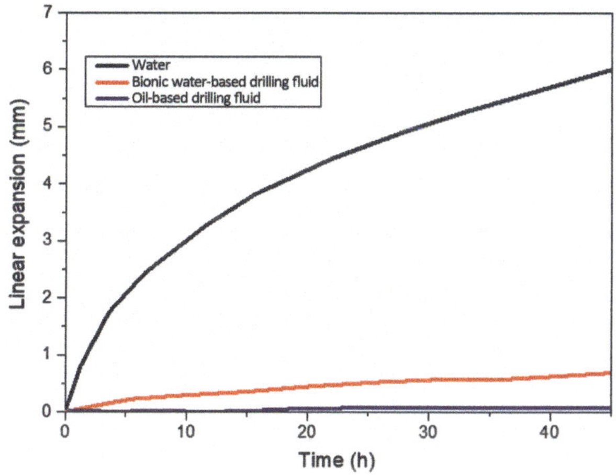

Figure 6.1 Comparison of the expansion height of bentonite with bionic water-based and oil-based drilling fluids.

Figure 6.2 Comparison of recovered cuttings.

Table 6.2 Comparison of cuttings recovery between bionic water-based and oil-based drilling fluids.

System	Recovery (%)
Pure water	3.69
Bionic water base	97.87
Oil base	99.27

3. Lubricity comparison

The comparison of lubricity experiments (Table 6.3) shows that the bionic water-based drilling fluid system had good lubricity, close to that of the oil-based drilling fluid.

Table 6.3 Lubricity comparison of bionic water-based and oil-based drilling fluids.

System	API filtration loss (mL)	Cake thickness (mm)	Sluggish coefficient
Bionic water base	0	1	0.037
Oil base	0.2	1	0.026

Table 6.4 Determination of heavy metal content in the bionic water-based drilling fluid.

Item	Result (mg/kg)	Maximum permissible content (mg/kg)
Cd	0	20
Hg	2.41×10^{-3}	15
Pb	17.2	1000
TCr	14.3	1000
As	0.62	75

Note: The heavy metal content was far below the standard value.

4. Evaluation of environmental protection

The environmental monitoring results show that the bionic drilling fluid system was nontoxic, safe, and environmentally friendly, with the results shown in Tables 6.4–6.6 and Fig. 6.3.

By changing the type and amount of bionic agents, the conventional water-based drilling fluid can be transformed into a bionic water-based drilling fluid. The following are several formulations of bionic high-efficacy water-based drilling fluids with different densities and temperature resistances, all of which are modified from different types of conventional water-based drilling fluids and have achieved the goal of high efficacy.

1. Density of 2.4 g/cm^3, temperature resistance of 120°C:

   0.1% Large potassium + 0.3% PAV-LV + 0.2% HEC + 2% starch + 2% plugging agent ZHFD + 1% film-forming agent + 1% bionic inhibitor + 2% $CaCO_3$ + 1% wellbore strengthener + 1.5% lubricant + 5% KCl + 0.5% superhydrophobic + barite.

2. Density of 1.5 g/cm^3, temperature resistance of 120°C:

   0.15% Bentonite mud + 0.15% large potassium + 0.4% PAC-LV + 0.2% HEC + 2% starch + 3% white asphalt + 2% film-forming agent + 0.3% bionic inhibitor + 0.5% superhydrophobic + 1.5% lubricant + 0.5% wellbore strengthener + 7% KCl + barite.

Table 6.5 Evaluation of the biological toxicity of the bionic water-based drilling fluid.

Sample	EC_{50} (mg/L)	Grade of biotoxicity
Mud of the bionic water-based drilling fluid	3.18×10^4	Nontoxic

Table 6.6 Biodegradability evaluation of the bionic water-based drilling fluid.

Sample	COD_{Cr} (mg/L)	BOD_5 (mg/L)	BOD_5/COD_{Cr}	Biodegradability
Mud of bionic water-based drilling fluid	1.32×10^5	2.54×10^4	0.192	Biodegradable

Figure 6.3 Vegetation recovery after discharge of the waste bionic drilling fluid system for one year.

3. Density of 1.4 g/cm^3, temperature of 120°C:
   2.5% Bentonite mud + 0.2% large potassium + 0.2% PAC-LV + 0.2% HEC + 3% starch + 0.5% highly effective anticollapse agent + 0.3% polyamino acid inhibitor + 3% ZHFD (plugging agent) + 4% KCl + barite.
4. Density of 2.2 g/cm^3, temperature of 150°C:

Soilless system: 0.1% KPAM + 0.3% PAC-LV + 1.2% E22 + 0.25% P-3S + 3% white asphalt + 2% $CaCO_3$ + 0.5% bionic inhibitor + 1.5% lubricant + 10% KCl + barite.

With soil phase system: 3% soil mud + 0.15% EB2 + 0.3% PAC-LV + 1.2% E22 + 0.33% P-3S + 3% white asphalt + 4% CaCO3 + 7% KCl + 30% NaCl + recrystallization stone.

In short, with the bionic wellbore strengthener, bionic inhibitor, bionic micro-nano plugging agent, bionic bonded lubricant, and amphiphobic agent as the core agents, bionic high-performance water-based drilling fluids can be prepared directly or by transforming other drilling fluids to meet different requirements.

6.1.2 New technology of chlorine-free high-efficacy, environmentally friendly bionic water-based drilling fluids

Inorganic salts such as KCl or NaCl have strong inhibitory properties and have become an indispensable component in almost all water-based drilling fluids. However, when waste drilling fluids containing these inorganic salts are discharged into the natural environment, the high Cl^- concentration will affect plant growth and human health, causing environmental pollution. Therefore taking advantage of the high inhibition capacity of the bionic drilling fluid agents, environmentally friendly drilling fluid systems without chloride salts were developed, including soil-containing systems and soil-free systems.

6.1.2.1 New technology of soil-phase, chloride-free, high-efficacy bionic water-based drilling fluids

The soil-containing, chloride-free high-efficacy bionic water-based drilling fluid systems developed by the research are as follows:

Formula 1: 1% bentonite mud + 0.2% bionic wellbore strengthener + 0.8% filtration control agent PAC-LV + 2% starch + 3% bionic plugging agent + 1.2% bionic inhibitor I + 0.5% bionic inhibitor II + barite's density adjusted to 1.4 g/cm^3.

Formula 2: Formula 1 + 10% bentonite

To obtain formula 2, 10% bentonite was added to formula 1, and the soil contamination resistance (i.e., the inhibition capacity) was evaluated, as shown in Table 6.7. It can be seen that formula 1 and formula 2 both exhibited good rheological properties, and the high temperature and high pressure (HTHP) filtration loss for both was only about 8 mL. The cakes are shown in Fig. 6.4.

At the same time, using the cuttings in China's Karamay, the rolling recovery of the cuttings by formula 1 was examined, as shown in Table 6.8. It can be seen that the inhibition ability of the system was

Table 6.7 Evaluation of chloride -free environmentally friendly drilling fluid and soil contamination resistance.

System	Condition	AV (mPa · s)	PV (mPa · s)	YP/Pa	Gel10″/10′ (Pa)	$\varphi 6/\varphi 3$	FL_{API} (mL)	FL_{HTHP} (mL)	pH
1	Before hot rolling	54	50	4	0.5/0.5	2/1	1.4	8.2	9
	After hot rolling for 16 h	32	32	0	0.5/0.5	1/0			
2	Before hot rolling	74	64	10	1/3.5	3/2	1.8	8.6	9
	After hot rolling for 16 h	52.5	47	5.5	0.5/0.5	2/1			

Note: Density of 1.4 g/cm^3, hot rolling condition of 120°C × 16 h. Abbreviations: *HTHP*, high temperature and high pressure.

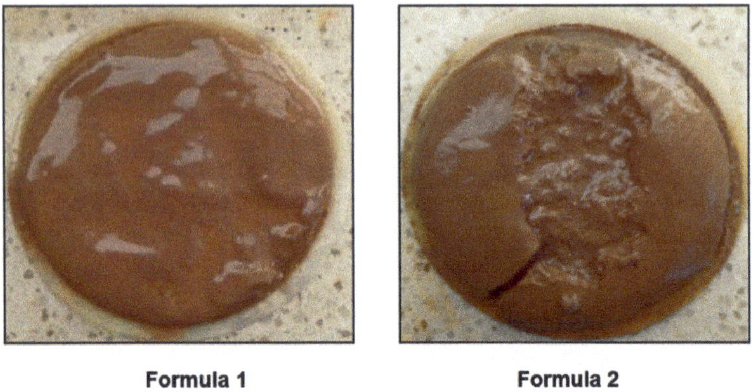

Figure 6.4 Morphology of the cakes.

Table 6.8 Recovery of chloride -free environmentally friendly drilling fluid system.

System	Hot rolling recovery (%)
Pure water	4.92
Formula 1	88.55

Note: Hot rolling condition of 120°C × 16 h.

Figure 6.5 Morphology of the hot rolled cuttings in the recovery experiment.

strong, and the recovery of cuttings increased by more than 94%. The hot rolled cuttings are shown in Fig. 6.5.

6.1.2.2 New technology of soilless, chloride-free, high-efficacy bionic water-based drilling fluids

The formula of high-efficacy bionic soilless and chloride-free water-based drilling fluid is 0.55% KPAM + 0.5% PAC-LV + 2% starch + 3% white asphalt + 1.1% bionic inhibitor + 0.5% (2,3-epoxypropyltrimethylammonium chloride) + 240 g barite.

Different amounts of bentonite were added to the system to investigate the soil contamination resistance of the system. The experimental results are shown in Table 6.9. The cakes are shown in Fig. 6.6.

When the amount of bentonite was increased from 0% to 15%, the rheological parameters and filtration loss of the system did not change much. After the amount of bentonite was increased to 20%, the viscosity of the system increased greatly, indicating that the system had good resistance to soil contamination.

It can be seen from Table 6.10 that the system had good inhibition, and the rolling cutting recovery was as high as 95% or more and was still more than 92% after two days.

6.1.3 One-trip drilling bionic high-efficacy water-based drilling fluid system

To achieve the goal of cost reduction and efficiency increase, the strong inhibition and lubricity of the bionic drilling fluid can be used to reduce the number of spuds and casings in drilling process, which is one of the main ways to reduce drilling costs and improve benefits. The following takes the Songnan shale oil formation in Jilin, China, as an example to demonstrate the design of an one-trip drilling bionic high-efficacy water-based drilling fluid system.

6.1.3.1 Evaluation of core agents

Owing to the large differences in rock characteristics and formation pressure coefficients of the Songnan shale oil formation, to reduce drilling costs, the original four-spud drilling was changed to three-spud drilling, and the original three-spud drilling was changed to two-spud drilling. This imposes more stringent requirements for drilling fluid performance. The existing drilling fluid systems cannot meet the requirements of formations in Songnan shale oil that collapse easily and leak easily. New drilling fluid technology needs to be developed to improve the drilling fluid's inhibition, anticollapse, and leak prevention performance; to achieve borehole stability; and to ensure the safe and complete drilling of the shale oil, thereby achieving the purpose of desired exploration. To achieve the goal of combining three-spud and four-spud drilling into one-spud drilling, the strengthening and inhibition of shale are the core issues. For this purpose, bionic wellbore strengtheners and bionic inhibitors are selected, and the following performances are evaluated.

Table 6.9 Basic performance test of the systems.

System	Condition	Φ6/Φ3	Initial/final gel strength (Pa/Pa)	AV (mPa · s)	PV (mPa · s)	YP (Pa)	YP/PV/Pa (mPa · s)	FL_{HTHP} (mL)
Formula	Before hot rolling	4/2	1/2	53.5	38	15.5	0.290	—
	After hot rolling	1/0	0.5/0.5	42	32	10	0.238	7.2
Formula + 5% bentonite	Before hot rolling	4/2	1/4.5	59	38	21	0.356	
	After hot rolling	2/1	0.5/1	51.5	41	10.5	0.204	8.4
Formula + 10% bentonite	Before hot rolling	5/3	1.5/4.5	77	58	19	0.247	—
	After hot rolling	3/2	1/1.5	58	44	14	0.241	7.2
Formula + 15% bentonite	Before hot rolling	10/6	2.5/8.5	88.5	59	29.5	0.333	
	After hot rolling	4/3	1/2	79.5	64	15.5	0.195	7.2
Formula + 20% bentonite	Before hot rolling	15/11	/19.5	131	87	44	0.336	
	After hot rolling	9/6	3/10	127	97	30	0.236	7

Note: Hot rolling temperature of 120°C × 16 h, HTHP temperature of 120°C, addition of back pressure, pressure differential of 3.5 MPa, density of 1.4 g/cm^3.
Abbreviations: *HTHP*, high temperature and high pressure.

Figure 6.6 HTHP cake. *Left*: formula 2 + 5% bentonite. *Middle*: formula 2 + 10% bentonite. *Right*: formula 2 + 15% bentonite. Abbreviations: *HTHP*, high temperature and high pressure.

Table 6.10 Hot rolling recovery of the cuttings.

System	Mass of hot rolled cuttings (g)	Hot rolling recovery (%)
Pure water	16 h	27.05
Bionic drilling fluid system	24 h	95.1
	48 h	92.5

1. Performance evaluation of the bionic wellbore strengthener

    To improve the anticollapse performance of Songnan shale oil formation, a bionic wellbore strengthener was used to generate a strengthening film on the surface of the shale, thereby enhancing the borehole stability. The Songnan shale's immersion experiment shows that when Jilin cuttings were added to pure water, a large number of cuttings fell off. The cuttings remained intact after the wellbore strengthener was added to the pure water, indicating that the wellbore strengthener had a good inhibitory effect on Jilin's hard and brittle cuttings. The results are shown in Fig. 6.7.

    The shale's inhibitory performance of the bionic wellbore strengthener GBFS-1 was evaluated through hot rolling recovery experiments on the shale cuttings of Jilin Oilfield. The hot rolling recovery of shale cuttings in different concentrations of GBFS-1 solutions is shown in

Figure 6.7 Changes in the morphology of shale cuttings immersed in different solutions.

Table 6.11 Hot rolling recovery of shale cuttings in the bionic wellbore strengthener solution.

GBFS-1 concentration (%)	Initial mass (g)	Recovered mass (g)	Hot rolling recovery (%)
0	30.0	5.48	18.27
0.1	30.0	16.95	56.5
0.3	30.0	23.61	78.7
0.5	30.0	27.87	92.9
1.0	30.0	28.86	96.2

Note: Hot rolling condition of 100°C × 16 h.

Table 6.11. The recovery of shale in pure water was 18.27%, and the addition of only 0.1% concentration of the GBFS-1 solution greatly increased the recovery to 56.5%. With the increase of the GBFS-1 concentration to 1%, the recovery of cuttings increased to 96.2%. This shows that the bionic wellbore strengthener had an excellent ability to prevent hydration and dispersion of shale.

2. Performance evaluation of the bionic inhibitor

Table 6.12 presents the results of the hot rolling recovery experiment of shale in pure water and different inhibitor solutions. After the hot rolling of shale cuttings at 100°C for 16 h in pure water, the recovery was only 18.27%, while the recovery in the 1% bionic inhibitor YZFS-1 solution reached 74.03%, which was significantly higher than that of other inhibitor solutions at the same concentration. With the increase of YZFS-1 concentration, the hot rolling recovery was further improved, which proved the excellent shale inhibitory performance of the bionic shale inhibitor YZFS-1.

6.1.3.2 Formula and performance evaluation of one-trip drilling bionic drilling fluid

6.1.3.2.1 Formula design

Considering the actual situation and cost, the basic idea is to transform the second-spud drilling fluid system into a one-trip drilling bionic drilling fluid.

Original formula (formula 1): 4% bentonite + 0.3% soda ash + 1% polyacrylonitrile-polyacrylamide composite ammonium salt for drilling fluid + 0.2% polyacrylamide potassium salt KPAM for drilling fluid + 1% high-efficacy silanol inhibitor DS-302 for drilling fluid + 1% polyester asphalt plugging agent JJFT-1 for drilling fluid + 1% low-fluorescent asphalt copolymer wellbore strengthener HQ-1 for drilling fluid + 3% white oil + 1% emulsifier LHR-II for drilling fluid.

Table 6.12 Hot rolling recovery of shale cuttings in pure water and different inhibitor solutions with concentration of 1%.

Inhibitor	Initial mass (g)	Recovered mass (g)	Hot rolling recovery (%)
Pure water	30.0	5.48	18.27
0.2% YZFS-1 solution	30.00	16.07	53.57
1% YZFS-1 solution	30.00	22.21	74.03
1.5% YZFS-1 solution	30.00	23.18	77.27
1% NW-1 solution	30.00	8.72	29.07
1% KCl solution	30.00	10.21	34.03
1% Ultrahib solution	30.00	23.29	73.63

Formula 2: Formula 1 + 0.5% wellbore strengthener.

Formula 3: Formula 1 + 1% wellbore strengthener.

Formula 4: Formula 1 + 2% wellbore strengthener.

Formula 5: Formula 1 + 2% wellbore strengthener + 1% inhibitor.

The measured rheological properties and filtration loss properties of these five formulas are shown in Table 6.13. The following can be seen:

1. When the wellbore strengthener was added to the original mud, the viscosity of the system did not change much, and the filtration loss of the system decreased.
2. With the increase in the wellbore strengthener concentration, the viscosity of the system decreased slightly, and the filtration loss did not change much.
3. Formula 5 was selected as the formula for the one-trip drilling bionic drilling fluid.

6.1.3.2.2 Performance evaluation

1. Evaluation of salt resistance

   To investigate the salt resistance of the system, 2%, 4%, 6%, and 8% NaCl were added to the system. The experimental results are shown in Table 6.14. The system maintained good rheological properties and filtration loss under 8% NaCl contamination, indicating that the system had good salt resistance.
2. Evaluation of anticalcium performance

   To investigate the calcium resistance of the system, 0.2%, 0.4%, and 0.6% of $CaCl_2$ were added to the system. The experimental results are shown in Table 6.15. The system maintained good rheological properties and filtration loss under 0.6% $CaCl_2$ contamination, indicating that the system had good calcium resistance.
3. Evaluation of inhibitory performance

   To investigate the effect of the system on borehole stability, the cuttings from Jilin were used for hot rolling recovery experiments and compared with the original system. The experimental steps were as follows: 20 g of cuttings (6–10 mesh) were prepared and hot rolled at 100°C for 16 h. The cuttings were then washed with water, dried, and weighed.

Table 6.13 On-site performance of the drilling fluids.

Formula	Condition before or after hot rolling	ρ (g/cm^3)	AV (mPa·s)	PV (mPa·s)	YP (Pa)	YP/PV	GEL10″/10′ (Pa)	FL_{API} (mL)	FL_{HTHP} (mL)	pH
1	Before hot rolling	1.28	26	19	7	0.368	4/7	—	—	9
	After hot rolling	1.28	27	20	7	0.35	3/6	2.8	9.4	8
2	Before hot rolling	1.28	29	21	8	0.38	4/8	—	—	9
	After hot rolling	1.28	27	20	7	0.35	3/6.5	2.4	9.0	8
3	Before hot rolling	1.28	30	22	8	0.36	4/8.5	—	—	9
	After hot rolling	1.28	28	21	7	0.33	3.5/6.5	2.2	8.8	8
4	Before hot rolling	1.28	33	24	9	0.375	5/9	—	—	9
	After hot rolling	1.28	31	23	8	0.348	4/7	2.0	8.4	8
5	Before hot rolling	1.28	32	24	8	0.333	4.5/9	—	—	9
	After hot rolling	1.28	29	22	7	0.318	4/6.5	2.2	8.6	8

Note: Hot rolling condition of 100°C × 16 h, pressure differential of 3.5 MPa. Abbreviations: *HTHP*, high temperature and high pressure.

Table 6.14 Evaluation results of various systems after hot rolling (100°C aging 16 h).

Salt content (%)	GEL10″/10′ (Pa)	AV (mPa · s)	PV (mPa · s)	YP (Pa)	YP/PV	FL_{API} (mL)	FL_{HTHP} (mL)	ρ (g/cm^3)	pH
0	4/6.5	29	22	7	0.318	2.2	8.6	1.28	9
2	4/6.5	27.5	21	6.5	0.310	2.2	8.6	1.28	9
4	4/6.5	27.5	21	6.5	0.310	2.4	8.8	1.28	9
6	3.5/6	26	20	6	0.3	2.4	9.0	1.28	9
8	3.5/6	24.5	19	5.5	0.289	2.6	9.2	1.28	9

Note: Hot rolling condition of 100°C × 16 h, pressure differential of 3.5 MPa. Abbreviations: *HTHP*, high temperature and high pressure.

Table 6.15 Evaluation results of various systems after hot rolling (100°C, 16 h).

$CaCl_2$ content (%)	GEL10″/10′ (Pa)	AV (mPa · s)	PV (mPa · s)	YP (Pa)	YP/PV	FL_{API} (mL)	FL_{HTHP} (mL)	ρ (g/cm^3)	pH
0	4/6.5	29	22	7	0.318	2.2	8.6	**1.28**	9
0.2	3.5/6	26	20	6	0.3	2.4	9.0	1.28	9
0.4	3.5/6	24.5	19	5.5	0.289	2.8	9.4	1.28	9
0.6	3/5.5	23	18	5	0.278	3	9.6	1.28	9

Note: Hot rolling condition of 100°C × 16 h, pressure differential of 3.5 MPa. Abbreviations: *HTHP*, high temperature and high pressure.

Table 6.16 Hot rolling recovery experiments of cuttings.

Formula	Mass before hot rolling (g)	Mass after hot rolling (g)	Recovery (%)
1	20.00	14.58	72.9
2	20.00	18.86	94.3

Note: Hot rolling condition of 100°C × 16 h.

The results are shown in Table 6.16. When the inhibitor and wellbore strengtheners were added to the system, the recovery of the system was more than 94%, which was over 20% higher than that of the original system. Therefore the system had a strong inhibitory performance.

To further investigate the effect of the system on borehole stability, the experiment of linearly expanding the cuttings in Jilin was conducted to compare the anticollapse system with the original system. The results are shown in Fig. 6.8 and Table 6.17.

It can be seen that when the inhibitor and wellbore strengthener were added to the system, the rate of linear expansion decrease of the system reached 77.40%, which was over 20% higher than that of the original system. Therefore the system had a strong inhibitory effect.

4. Evaluation of compressive strength

Through uniaxial stress tests on the cores immersed in the drilling fluid systems, the strengthening effect of the bionic drilling fluid on shale was investigated. The two systems were left standing for 16 h at 100°C. The uniaxial failure experiments were conducted on cores 1 and 2 on the TAW-1000 rock three-axis servo tester to evaluate the strengthening of the cores that had been treated by the bionic drilling fluid. The experimental uniaxial stress-strain curve is shown in Table 6.18. The core failure strength of the system with the bionic

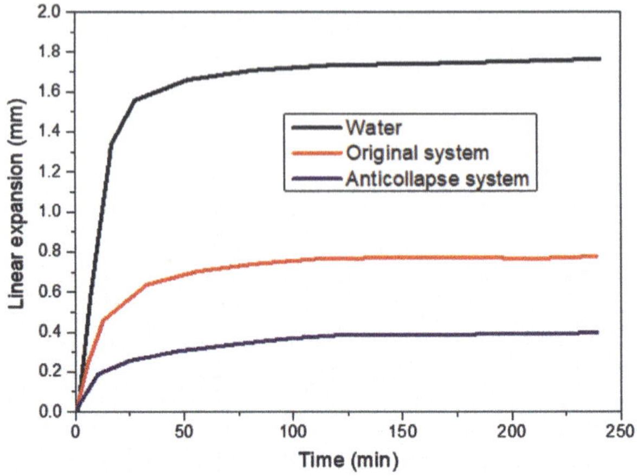

Figure 6.8 Linear expansion of cuttings powder.

Table 6.17 Linear expansion test of cuttings powder.

System	Expansion height (mm)	Rate of expansion decrease (%)
Pure	1.77	—
Original drilling fluid system	0.78	55.93
One-trip bionic water-based drilling fluid system	0.4	77.40

wellbore strengthener was much greater than that of the system without strengthening. Therefore the bionic anticollapse drilling fluid system had a better effect of strengthening the borehole.

5. Evaluation of self-plugging performance of cuttings

For formations that easily collapse and leak, it is necessary to add solid particles to the drilling fluid to plug the pores. During the drilling process, there are a large number of solid particles of different sizes in the drilling fluid system (cuttings, weighting materials, etc.). Therefore the team led by Professor Guancheng Jiang of China University of Petroleum (Beijing) proposed using the inherent solid-phase particles in the drilling fluid system for the first time to prevent the leakage of drilling fluid and to prevent pressure transmission and save costs. To evaluate the feasibility and effect of this idea, 3% cuttings (60–80 mesh) were added to the one-trip bionic water-based drilling fluid, and the closure plate was used to evaluate the effectiveness of the self-plugging.

Table 6.18 Conventional mechanical parameters of cores.

Formula	Diameter (mm)	Breaking strength (MPa)	Increase rate (%)
1	25	4.72	–
2	25	5.22	10.59

Table 6.19 Evaluation of self-plugging performance of one-trip bionic drilling fluid system.

Time (min)	5	10	20	30	40	50	60
Filtration loss on sand bed (mL)	0	0	0	0	0	0	0

Note: Experimental temperature of 100°C and pressure differential of 3.5 MPa.

Figure 6.9 On-site cuttings (well depth of 100–1200 m).

Evaluation method: 100 g of 20–40 mesh and 40–70 mesh quartz sand was added to the mud cup of the Model 71 HTHP water loss instrument, compacted, and flattened. The above drilling fluid was slowly poured onto the sand bed after hot rolling of 100°C for 16 h. The filtration loss of the drilling fluid on the sand bed in 1 h was recorded, and the self-plugging performance was evaluated according to the filtration loss. The results are shown in Table 6.19. The one-trip bionic drilling fluid had zero filtration loss on the 40–70 mesh sand bed, which could self-plug the pores and fractures of the formation rocks (Fig. 6.9).

Next, 80 g of 20–40 mesh sand and 20 g 40–70 mesh sand were added to the filter loss instrument. After compaction and flattening, the hot-rolled HTHP drilling fluid (300 mL) was added, the starting pressure was 2 MPa, and the pressure was slowly increased to 3.5 MPa after no

Table 6.20 Evaluation of cuttings' self-plugging performance of the one-trip bionic drilling fluid system.

Parameter	Formula	2 MPa	2–3 MPa	3 MPa (30 min)
Filtration loss (MmL)	Drilling fluid system	53	53	53
	Drilling fluid system + 3% cuttings	24	24	24

Note: Experiment temperature of 100°C and pressure differential of 3.5 MPa.

leakage. The filtration loss was measured after stabilization for 30 min, as shown in Table 6.20. The one-trip bionic drilling fluid system could use formation cuttings (in fact, the cuttings and weighting materials in the drilling fluid were distributed within the particle size range of the added sand) for self-plugging and good cakes were formed. Under a constant pressure of 3 MPa, the filtration loss of the drilling fluid was 24 mL, which was much smaller than the blank drilling fluid system (53 mL). It can be seen that the drilling fluid could form effective plugging in the sand bed and form strong cakes to bear certain pressure.

6.2 Field application

The new high-efficacy, low-cost, and environmentally friendly bionic water-based drilling fluid technology formed with a series of bionic drilling fluid agents as the core has been tested and applied in the oil fields of Xinjiang, North China, Changqing, Jilin, and Shengli. This technology has also been adopted by Schlumberger Company, and good application results have been achieved. This technology has now become the main technology of environmentally friendly drilling fluids for complex wells.

6.2.1 Application in Xinjiang Oilfield

In 2017 the CNPC Xibu Drilling Engineering Company conducted field applications in 11 wells in different areas of Xinjiang Oilfield and achieved good results, as shown in Figs. 6.10–6.14, demonstrating the following advantages of this technology: (1) Compared with other advanced technologies used in the adjacent wells in the same area, the drilling fluid had more stable performance, showed better inhibition and lubricity, was equivalent to oil-based drilling fluid, and had good environmentally

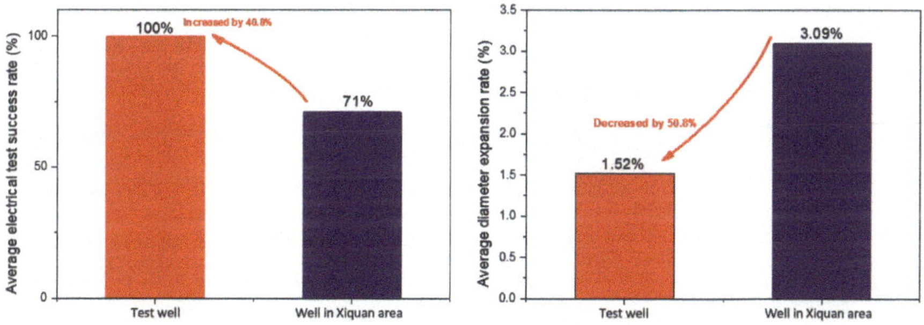

Figure 6.10 Comparison of the success rate of electrical survey and the borehole diameter expansion rate of the application wells and the compared wells in the Xiquan area, Xinjiang.

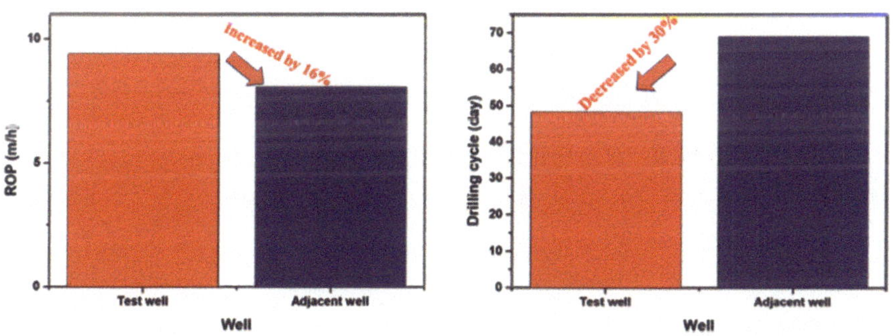

Figure 6.11 Comparison of ROP and drilling cycle of application wells and compared wells in Xiquan area, Xinjiang.

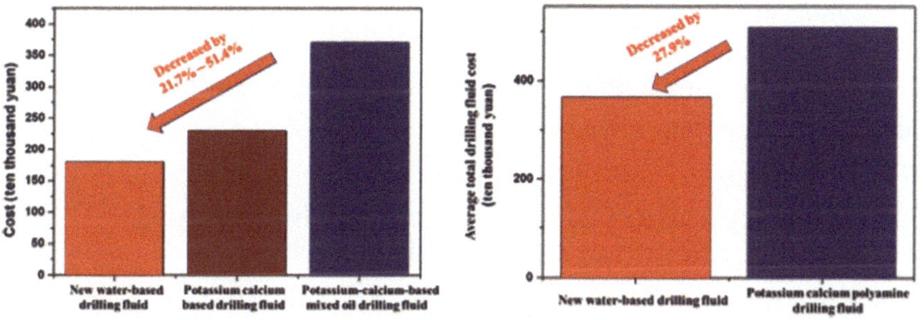

Figure 6.12 Comparison of costs between the developed technology and other drilling fluid technologies in Xiquan (*left*) and Mahu (*right*) areas, Xinjiang.

friendly properties. (2) The success rate of the electrical survey was 100% (increased by 40.8%). (3) The number of complicated conditions downhole was reduced by 81.9%, the penetration rate increased by more than

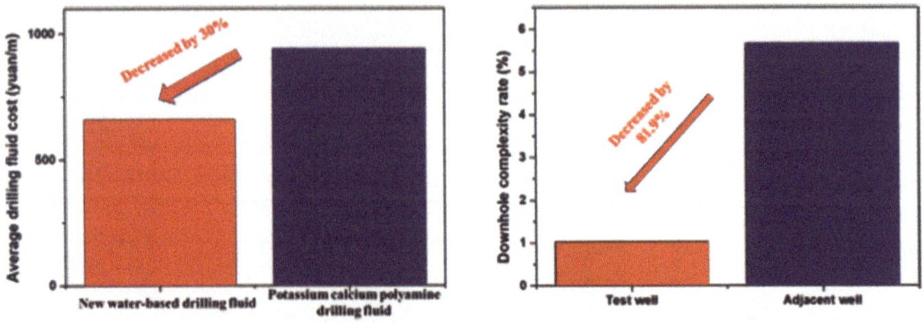

Figure 6.13 Comparison of drilling fluid cost and downhole complex situations between applied wells and compared wells in Mahu area, Xinjiang.

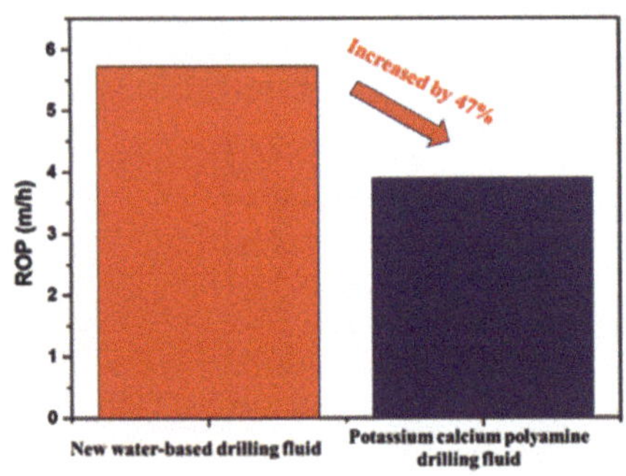

Figure 6.14 Comparison of ROP between application wells and compared wells in Mahu area, Xinjiang.

16% on average, and the drilling cycle was reduced by 30% on average. (4) Drilling fluid costs were reduced by 30% on average. This technology has become a water-based drilling fluid technology with Chinese characteristics and has provided solutions to technical problems that had previously been difficult to solve.

6.2.2 The application of one-trip bionic drilling fluid technology in Songnan shale of Jilin Province

Songnan shale oil is a new exploration field in the deep part of the Xinbei complex block of Jilin Oilfield, with shallow burial depth, high oil saturation, and rich resources. Upper Heidimiao shallow gas is active, there are many injection and production wells, the wellhead pressure is high, and the well control risk is high. Formation pressure is chaotic, drilling accidents are frequent, and there are wells that cannot be successfully completed and have been discarded. The formation pressure of newly explored targets is low, and the development of fractures and the existence of multiple pressure systems throughout the well have brought great challenges to safe drilling. For example, the formation pressure of well Xin 373 is high in the upper part and low in the lower part; the formation is easy to deviate; cracks develop; the problems of collapse, block loss, and lost circulation are serious; the safety density window is narrow; and well cementing faces many difficulties. The following problems are encountered in the drilling process:

1. The shallow gas pay in the Heidimiao formation is active, and the water injection pressure of adjacent wells is high.

   The lithology of Heidimiao formation is dominated by shale that has good drillability. The gas measurement value is high, the formation pressure is high, water intrusion occurs during drilling, and the drilling fluid density required to balance the formation reaches more than 1.80 g/cm^3.
2. The Putaohua formation has serious problems of lost circulation and block falling, and the stratum is easily deviated.
   a. The Yaojia formation has high gas measurement value during drilling, the highest total hydrocarbon being 100%, and good showing. When the drilling fluid density reaches 1.35–1.50 g/cm^3, overflow and leakage occur simultaneously. When the drilling fluid density reaches 1.54 g/cm^3, the filtration loss increases.
   b. After drilling to 800 m, serious pipe sticking occurs as well as the problems of collapse, block falling, and pump blockage.
   c. The formation is easy to deviate, and it is difficult to control the hole drift angle.
3. The shale fractures in the Gaotaizi layer are developed, and the collapse and block falling are serious.
   a. The Qingshankou formation has vertical fractures, the blocks tend to fall off during drilling, and the formation is easy to peel off when pigging the well.
   b. The well pigging encounters obstacle and requires, frequent redressing, and it is difficult to remove the obstacles.
4. There are also problems of poor well conditions, lost circulation, small wellbore clearance, and difficulty in well cementing.

   To ensure the safe drilling and completion of wells, even the previous method of four-spud drilling could not solve the problems. By adopting the high-efficacy bionic water-based drilling fluid technology developed by the team of Professor Guancheng Jiang from China University of Petroleum (Beijing), safe drilling and completion with the conventional three-spud structure have been realized, and normal construction has been ensured, such as drilling coring and completion logging.

Well Xin 371 is in the northern part of the Xinli structure in the Fuxin uplift in the Songnan central depression. The purpose of drilling is to reveal the exploration potential of fractured shale reservoirs in the Qing III section of the Xinbei area. This well is an exploratory well in the Jilin Oilfield with a well depth of 1400 m. The target layer is the Qing II section, and the Gaotaizi oil formation of Qing III section is drilled through.

The one-trip bionic drilling fluid technology has been successfully applied in well Xin 371. Under the condition that the upper and lower formation pressures are quite different and there is a risk of well collapse and lost circulation, the upper and lower formations were well transformed from two-spud drilling and completion into one-spud drilling and completion. The drilling process was safe, no downhole complications occurred, and drilling costs were saved. At the same time, the drilling fluid system has the following characteristics:

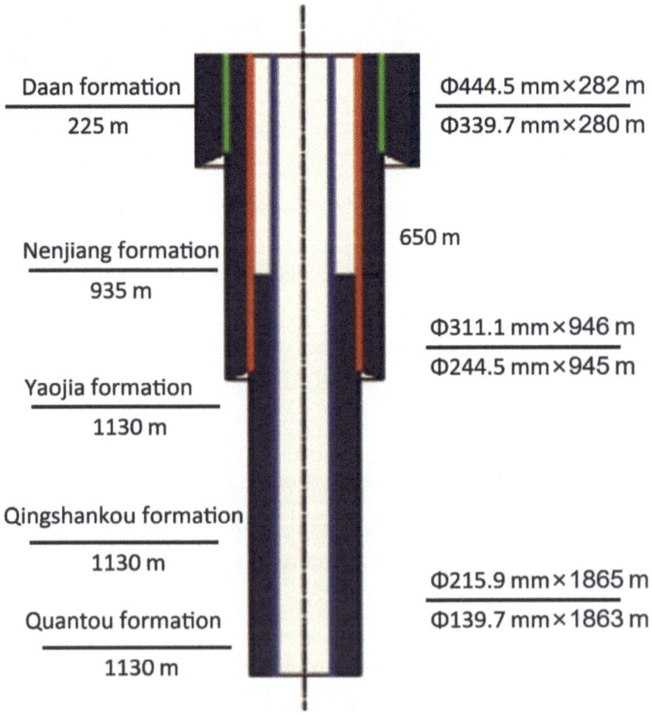

Figure 6.15 Schematic diagram of well Xin 380.

a. The wellbore strengthener could effectively improve the quality of the cake and enhance the stability of the borehole.
b. By adding wellbore strengthener, the rock could be cemented well, the borehole was stabilized, and the block falling could be significantly reduced.
c. There were no resistance or pipe sticking during the electrical test, and the process of the electrical test was smooth.

5. Application in well Xin 380

Well Xin 380 is in the northern part of the Xinli structure in the Fuxin uplift belt in the central depression area of the southern Songliao Basin. The purpose of drilling was to reveal the exploration potential of fractured shale reservoirs in the Qing III section of the Xinbei area. The well was 1865.00 m deep and was an exploratory well. The target layer was the Gaotaizi oil formation of Qing III section. In the presence of major difficulties in this well, three-spud drilling and completions were still designed (Fig. 6.15).

Under harsh conditions the drilling fluid system well dealt with the contradiction between the upper and lower strata and achieved the following effects:

1. When the amount of bionic wellbore strengthener was 1%, the filtration loss of the drilling fluid did not increase, but the quality of the cake was significantly improved.
2. After the bionic wellbore strengthener was added, the cuttings returned by the vibrating screen were sharply angular and dry inside, which improved the antiencapsulating performance of the drilling fluid.
3. The process of drilling coring and electrical measurement was smoothly conducted.

References

[1] R.G. Bland, R.R. Waughman, P.G. Tomkins, et al., Water-based alternatives to oil-based muds: do they actually exist? SPE 74542, 2002.
[2] J. Friedheim, G. Sartor, New water-base drilling fluid makes mark in GOM, Drilling Contractor, 2002.
[3] W. Dye, K. Daugereau, N. Hansen, et al., New water-based mud balances high-performance drilling and environmental compliance, SPE 92367, 2003.
[4] K. Morton, B. Bomar, M. Schiller, et al., Selection and e-valuation criteria for high performance drilling fluids, SPE 96342, 2005.
[5] Z. Qigen, C. Fu, L. Yi, et al., Development status of foreign high-performance water-based drilling fluid technology, Drill. Fluid Completion Fluid 24 (3) (2007) 74−77.
[6] S.E. Alford, A. Asko, M. Campbell, et al., Silicate-based fluid, mud recovery system combine to stabilize surface formations of azeri wells, SPE 92769, 2005.

[7] K. Dennis, B. Hugehes, et al., Advances high performance water-based drilling fluid design, in: CADE/CAODC Drilling Conference, October 23–24, 2001.
[8] W. Jianhua, Y. Jienian, D. Tongwei, Research progress of high-performance water-based drilling fluids, Drill. Completion Fluids 24 (1) (2007) 71–75.
[9] W. Dye, K. Augereau, N. Hansen, et al., New water-base mud balances high - performance drilling and environmental compliance, SPE 92367, 2004.
[10] E. Van Oort, Physico-chemical stabilization of shales, SPE 37263, 1997.
[11] T.W. Beihoffer, F.B. Growcock, D.S. Dorrough, et al., Field testing of a cationic polymer/brine drilling fluid in the northsea, SPE 24588, 1992.
[12] J.R. Bruton, H.C. Mel-aurine, Modified poly-amino acid hydration suppressant proves successful in controlling reactive shales, SPE 26237, 1993.
[13] Z. Zengfu, L. Ping, M. Shiqing, et al., Research and application of small cation-polyol drilling fluid technology, Drill. Fluid Completion Fluid 22 (2) (2005) 47–49.
[14] L. Wei, C. Zhixue, S. Mengci, et al., Research and application of drilling fluid technology for complex deep wells in Qingxi area, Drill. Fluid Completion Fluid 23 (3) (2006) 47–50.
[15] P. Jianzhong, W. Shuyong, L. Wenming, Solve the difficult problem of rheological adjustment of high temperature drilling fluid with organic amine treatment agent, Drill. Fluid Completion Fluid 26 (3) (2009) 79–81.
[16] G. Nana, Development and Performance Study of Drilling Fluid Inhibitors, Southwest Petroleum University, 2014.
[17] L. Huanyu, L. Chuanbei, Z. Dehua, et al., Application of high-performance water-based drilling fluid in Well Hong 69, Drill. Fluid Completion Fluid 28 (2) (2011) 87–89.
[18] X. Xianguo, Z. Qi'an, Research on a new type of amine-based polyol anti-collapse agent, Drill. Technol. 33 (1) (2009) 93–96.
[19] W. Jian, Z. Rong, N. Mingshun, et al., Research and application of HRD weak gel drilling fluid, Nat. Gas Explor. Dev. 31 (4) (2008) 52–56.
[20] H. Hongxi, X. Mingbiao, W. Changjun, et al., Laboratory study on the performance of weak gel non-solid polyamine drilling fluid, Oilfield Chem. 26 (1) (2009) 5–7.
[21] W. Changjun, X. Mingbiao, M. Hailong, Laboratory study of polyamine UHIB strong inhibitory drilling fluid, J. Oil Gas Technol. 31 (1) (2009) 80–84.
[22] M. Hailong, L. Guozhao, W. Changjun, Research and application of GID drilling fluid in Bozhong Oilfield, Drill. Fluid Completion Fluid 26 (2) (2009) 90–92.
[23] Y. Kezhi, X. Mingbiao, Application of polyamine drilling fluid in Liuhua 26-1-1 well in South China Sea, J. Oil Gas Technol. 33 (9) (2011) 119–123.
[24] X. Mingbiao, M. Shuangzheng, H. Jinfang, et al., Research on high temperature resistant solid-free weak gel drilling fluid system, Oilfield Chem. 29 (2) (2012) 142–145.
[25] J. Guancheng, M. Yuncai, Z. Baoyi, S. Ranran, Research progress and development trend of drilling fluid technology for protection of oil and gas formations with temporary plugging, Drill. Fluid Completion Fluid 35 (2) (2018) 1–16.
[26] Q. Yuanzhi, Development and evaluation of shale inhibitor SIAT, Pet. Drill. Technol. 37 (6) (2009) 53–57.
[27] Q. Yuanzhi, R. Kesheng, H. Hongjun, et al., Application of amine-based drilling fluid in Mo 116 well block of Xinjiang Oilfield, Drill. Fluid Completion Fluid 28 (6) (2011) 24–27.
[28] Q. Yuanzhi, Research on new amine-based inhibitors (I)-molecular structure design and synthesis, Drill. Fluid Completion Fluid 27 (1) (2010) 1–5.
[29] L. Tianke, Drilling fluid technology of Laixi 84 well, Drill. Fluid Completion Fluid 28 (2) (2011) 84–88.

CHAPTER 7

Development trends and prospects of bionic drilling fluid technologies

The previous chapters discussed how to introduce bionics into the field of drilling fluids and the relevant theories, methods, and technologies that can be used to solve the technical problems encountered in drilling fluids in the drilling process, such as borehole instability, pipe sticking, formation damage, lost circulation, and difficult hole cleaning. For example, we have developed corresponding new bionic drilling fluid materials and technologies, that is, a new technology of bionic borehole-enhanced drilling fluids, a new technology of bionic flow and bonded drilling fluids for preventing pipe sticking, a new technology of bionic drilling fluids to protect formations with different permeabilities, a new technology of bionic strong-adhesion high-pressure-bearing drilling fluids for intelligent leakage prevention and plugging, and a new technology of bionic drilling fluids featured by bubble wrapping and cuttings carrying. Not only has a bionic high-efficacy water-based drilling fluid been developed that features safety, high efficiency, low cost, and environmental friendliness, but also the drilling fluid has been well verified and promoted in the drilling practices of hundreds of wells in different formations in the world.

Although the new bionic drilling fluid technology has been established by the China University of Petroleum (Beijing) as a result of more than 10 years of uninterrupted theoretical research, indoor experiments, field tests, indoor improvement and on-site retesting, this is the first time in the world that the theories and methods of bionics have been adopted to establish a new theory and new technology of drilling fluids. As the formation conditions that are encountered become more and more complicated and harsh, many aspects of drilling technology need to be further improved. At the same time, the research methods of bionics and bionic materials are constantly being updated, and new research results are constantly emerging, providing new sources for in-depth research on bionic drilling fluids. This chapter explains the relationship between drilling fluid

Fundamentals and Applications of Bionic Drilling Fluids
DOI: https://doi.org/10.1016/B978-0-323-90293-9.00007-1

technology and drilling fluid materials as well as the main challenges of current drilling fluid technology and then discusses the development prospects of bionic drilling fluid materials and technologies, thereby providing some useful enlightenment for the further research and application of bionic drilling fluid materials.

7.1 Materials and technologies of water-based drilling fluids

The development of drilling fluid technology has gone through the starting stage in the 1960s, the development stage from the 1980s to the 1990s, and the improvement and in-depth development stages from the late 1990s to the beginning of the 21st century; it is now in the in-depth improvement stage because of the establishment of bionic drilling fluid technology. Thus it can be easily concluded that the development of drilling fluid technology is based on the development of drilling fluid agents or new materials. This section briefly introduces the development of different types of drilling fluid materials and technologies.

7.1.1 The starting stage of polymer drilling fluid, the proposal of nondispersed low solid polymer water-based drilling fluids

In the 1960s, petroleum engineering scientists and technicians gradually realized that the type, composition, and performance of drilling fluids directly affects the penetration rate and drilling cost. In particular, the solid content in a drilling fluid is the main factor affecting the penetration rate and drilling cost. After many years of research, agents such as flocculants, encapsulating agents, and inhibitors have been developed to prevent the solid particles in the drilling fluid from further dispersing and becoming finer and to inhibit the fine particles from becoming flocculated and agglomerated. Then with the use of surface solid control equipment, the solid particles, especially the useless solid phase in drilling fluids, are removed, which can increase the penetration rate and reduce the drilling fluid cost. For example, in 1966, Pan American Petroleum Company used this technology for the first time in the oilfields of western Canada, greatly increasing the penetration rate and reducing the direct cost by 2.5%. In the 1970s the penetration rate was increased by 17%– 28% in the Delawan Basin oilfield in the United States, and the drilling cost dropped by more than 20%. In 1973 the technology was successfully used for the first time in Shengli Oilfield, China, and with further promotion and application, it increased the penetration rate by more than 20%. Subsequently, scientific and technological workers have developed a variety of polymer drilling fluid agents and gradually formed two types of nondispersed low solid polymer drilling fluid technologies represented by partially hydrolyzed polyacrylamide and its derivatives and the copolymers of vinyl acetate and maleic anhydride, of which the partially hydrolyzed polyacrylamide drilling fluid system is still used in major oilfields in China, guaranteeing safe and efficient drilling.

7.1.2 The development stage of polymer drilling fluids: The development of nondispersed low-solid polymer water-based drilling fluids

Because the polymer drilling fluids described above have only a small variety of agents, cannot meet the needs of complex formations and deep wells above 3000 m, and cannot be easily controlled in terms of density, researchers have successfully developed some drilling fluid agents, including the 80A series (e.g., 80A51, 80A46, 80A40) [1–3], PAC series (e.g., PAC141, PAC142, PAC143) [1,2,4], and SK series (e.g., SK-I, SK-II, SK-III) [1–3]. As a result, the main agents for polymer drilling fluids have developed from single partially hydrolyzed polyacrylamide and its derivatives to a variety of metal salt with, large, medium, and small molecular weight gradations (molecular weight varies from 10,000 to 4 million or more), different degree of hydrolysis (30%–60%), and multiple functions (dilution, flow pattern improvement, filtration control, anticollapse, temperature resistance, salt resistance, etc.). A variety of polymer drilling fluid systems that can be adapted to the needs of different regions and different formation conditions have been developed, and obvious effects have been observed in production that have improved the performance of polymer drilling fluids. However, field applications have shown that the polymer water-based drilling fluid has shortcomings, such as excessive gel strength, poor cake quality, high viscosity under extreme high shear rate, and difficulty in meeting the needs of optimized drilling.

7.1.3 The improvement stage of polymer drilling fluids

To solve the contradictions between the optimization of rheological parameters and borehole stability, more than 10 years of hard work enabled scientific and technological workers to successfully develop cationic and zwitterionic polymer drilling fluids in the 1980s and early 1990s. Some of the contradictions have been solved to a certain extent.

7.1.3.1 Cationic polymer drilling fluids

Cationic polymer drilling fluid systems use small molecular clay stabilizers (also known as small cations) and high molecular clay encapsulating agents (also known as large cations) as core agents, supplemented by viscosifiers, filtration control agents, and plugging agents.

The hydrophobic groups of small cations are easily adsorbed on the surface of the clay to form a hydrophobic film that prevents the hydration expansion and dispersion of the clay particles. At the same time, the organic cation groups are inserted into the crystal layer to narrow the spacing between the crystal layers and prevent the entry of water molecules. In addition to hydrogen bonding and electrostatic attraction between the positively charged groups on the large cationic molecular chain and the clay particles, encapsulating and bridging can also occur, leading to strong encapsulating, flocculation, and inhibition capabilities. The synergistic effect of large and small cations can effectively inhibit clay hydration and cuttings dispersion and improve borehole stability. This drilling fluid technology has been successfully applied in the Erlian oilfield, for example Ref. [5].

Although cationic polymer drilling fluids have many advantages, the cationic properties can easily cause the clay particles to flocculate. At the same time, there are few supporting agents; poor compatibility with other anionic drilling fluid agents makes it difficult to adjust the system's rheological properties and causes high cost. Furthermore, the cationic polymer has high biological toxicity and does not meet environmental protection requirements, which limits its field application.

7.1.3.2 Zwitterionic polymer drilling fluid

To better realize optimization of the drilling process, researchers have developed a new type of zwitterionic polymer viscosity reducer (XY series, represented by XY-27), which reduces the structural viscosity (AV, YP, GS), reduces Casson limit viscosity rather than increasing it, and can also enhance the inhibitory ability of the drilling fluid system. This solves, to a certain extent, the difficulty of balancing viscosity reduction and inhibition.

With polymer viscosity reducer XY as the core agent, combined with a new type of strong encapsulating agent (FA series, represented by

FA-367), various types of zwitterionic polymer drilling fluid systems have been developed (e.g., solid-free zwitterion brine drilling fluids, low-solid nondispersed zwitterionic drilling fluids, low-density oil-mixed zwitterionic drilling fluids, temporary plugging type zwitterionic completion fluids, and high-density zwitterionic brine systems) [6–9]. The wide applications in China's oilfields show that these drilling fluids have a positive effect on optimizing drilling and to a large extent solve the drilling fluid problem caused by the mud making of the formation. Good results have been achieved, and the use range of polymer drilling fluids has been expanded [10–14].

7.1.4 The redevelopment stage of polymer drilling fluids: High-performance polymer water-based drilling fluids

In the 21st century the geological conditions of newly discovered oil and gas fields are mostly very complicated, and the difficulty of drilling operations and the cost of oil and gas development have also increased sharply. The complications include shale wells with ultralong sections, ultradeep wells, high-temperature wells, high-pressure wells, high-steep structures, thick or very thick salt rock formations or salt gypsum formations, formation fluids with high acidity and high salinity, formation fluids with high-value metal salts, extended reach wells, and deep-water wells. In most cases there are complex situations, such as resistance to round trip, pipe sticking, low penetration rate, well collapse, lost circulation, and serious formation damage. Water-based drilling fluid technology cannot always be used to solve these problems [15–20], while environmental pollution and high cost make it difficult to use oil-based and synthetic-based drilling fluids with superior inhibition and lubricity widely. As a result, scientific and technological workers have improved the comprehensive performance of polymer water-based drilling fluids by developing high-performance drilling fluid agents, hoping to form a series of high-performance water-based drilling fluids with inhibition and lubricity close to those of oil-based drilling fluids and to meet the requirements of optimized drilling [21–23]. Representative high-performance water-based drilling fluids include the ULTRADRILL [24–28] and HydraGlyde [29] drilling fluids of MI SWACO Company, Pure-Bore drilling fluids of INNOSPECINC Company [30,31], PERFORMAX [32,33] and LATIDRILL [34–37] drilling fluids of Baker Hughes Company, HYDRO-GUADRTM [38] and PERFORMADRIL [39,40] drilling fluids of Halliburton Company, and the polyamine high-performance water-based drilling fluids [41,42], multifactor synergistic high-performance water-based drilling fluids [43–50], and organic salt drilling fluids [51–53] of China.

7.1.5 The reimprovement stage of polymer water-based drilling fluids: Bionic high-efficacy water-based drilling fluids

The series of high-performance water-based drilling fluids that have been developed have mostly aimed to solve the problems of osmotic hydration, fluid lubrication, and plugging of large pores and cracks. Although the inhibition and lubricity of water-based drilling fluids have been improved, they still fall behind oil-based drilling fluids in some aspects, and it is difficult to meet the requirements for drilling fluids in terms of complex geological and ground conditions, costs, and environmental protection [39,40,54,55]. To this end, Professor Guancheng Jiang of China University of Petroleum (Beijing) led the team members to innovatively introduce bionics into the field of drilling fluids. By imitating the principle of adhesion proteins secreted by marine mussels to cement rocks, a bionic wellbore strengthener and a bionic inhibitor were developed. By imitating the characteristics and structure of bacteria, a bionic micro-nano plugging agent was invented, and wellbore strengthening was achieved through the combination of internal and external factors, so the ability to maintain borehole stability of water-based drilling fluids is close to that of oil-based drilling fluids, solving the technical problem of borehole instability. By imitating the principle of earthworms' secretion of mucus to reduce frictional resistance, a bionic bonded lubricant was developed, and a new method and technology to reduce downhole friction with bionic flow and bonded synergy were developed, making the lubricity of water-based drilling fluids close to that of oil-based drilling fluids. By imitating water spiders' superhydrophobic fluff to effectively adhere to bubbles in the water to provide buoyancy, a cuttings-carrying agent was developed to improve the gas wettability of the cuttings surface, reduce the relative density of the cuttings, and improve the cuttings-carrying ability of the drilling fluid. At the same time, the bionic superamphiphobic agent and bionic superhydrophobic agent were invented, which further improved the effect of stabilizing the borehole and protecting the formations. In addition, by simulating the way in which royal water lily leaves are connected by multiple nodes into a vein grid structure with high strength, high toughness, and uniform spatial force and combining this with the multilayer composite structure and composition of shells, new material and technology of bionic pressure-bearing plugging for stop conditions were developed, in which the plugging agent was made of deformable fiber mesh, high-strength gel, and rigid particles. By simulating the principle by which mammalian blood uses platelets and fibrinogen to automatically stop bleeding, new material and technology of plugging for in-drilling conditions were developed in which the material was made of nanowhiskers, weak gels, and particles. Finally, the theories and technologies of bionic drilling fluids were created, which have been promoted and applied in many oilfields in the world to solve the problems of wellbore instability, pipe sticking, lost circulation, formation damage, and difficult hole cleaning that had been difficult to solve before. The comprehensive performance of polymer water-based drilling fluids have been further improved.

It can be seen that the advent of new drilling fluid technologies at all stages is based on the development of new drilling fluid materials. In other words, only by successfully developing new drilling fluid materials can new drilling fluid technology be developed, and the further development of bionic drilling fluid technology is no exception.

7.2 Development trends and prospects of bionic drilling fluid technologies

In recent years, most of the major newly discovered oil and gas resources in the world have been found in deep ocean waters and unconventional resources. The Arctic Circle also contains abundant oil and gas resources. In other words, most of the remaining oil and gas resources in the world are in difficult-to-explore places featuring low-permeability and unconventional, deep-water and deep-formation properties. They are mainly distributed in mountainous areas, deserts, plateaus, loess plateaus, and ocean-covered areas. So the geological and ground conditions that are encountered are becoming increasingly complex, and exploration and development are getting increasingly difficult, especially in the current situation of low oil prices and strict environmental protection requirements. As a result, the related technologies, especially drilling fluid technologies, are facing unprecedented challenges. The existing drilling fluid theories, methods, and technologies have gradually become unable to meet the needs of the new situation. Therefore we must search for original innovations, pursue novel theoretical methods, and achieve key technological breakthroughs in order to develop new high-performance drilling fluid materials to form the new technology of high-efficacy drilling fluids.

7.2.1 The main challenges faced by the current drilling fluid technology

7.2.1.1 Unable to be discharged directly due to serious environmental pollution

In the oil-drilling industry the pollutants produced by oil and gas drilling each year have caused long-term contamination of hundreds of thousands of acres of soil and water and ecosystem degradation in China through diffusion and migration, where more than 90% of the pollutants come from drilling fluids and drill cuttings. Although environmental protection measures are now being taken, such as not allowing drilling wastes to land on the ground, the environmental pollution caused by the drilling fluid system is still serious.

The environmentally friendly drilling fluid technologies that are currently used include polyol environmentally friendly drilling fluid technology, methyl glucoside environmentally friendly drilling fluid, silicate environmentally friendly drilling fluid, and synthetic-based environmentally friendly drilling fluid. Although the environmentally friendly drilling fluid technology has achieved good results, there are still some limitations, such as complex construction process, cumbersome maintenance, lack of environmentally friendly agents, inconvenient use, high construction costs, and poor economic benefits. Therefore environmentally friendly drilling fluid technology has not been well promoted and applied, and the damage to the environment by waste drilling fluid is still serious.

7.2.1.2 Many complex downhole accidents

With the continuous increase in the scale and number of downhole construction operations in recent years, especially as the geological conditions encountered in drilling have become increasingly complex, the probability of complex accidents during downhole operations has also risen sharply [56]. For example, on-site investigations and studies have shown that accidents caused by lost circulation are very frequent, accounting for over 58% of all accidents [56]. Second, well collapse and complex accidents account for approximately 24.5% of all accidents. In addition, obstacles and well surges are also common in the well logging process. These high frequencies of downhole complex accidents require the improvement of the comprehensive performance of drilling fluids.

1. Lost circulation

   So far, although there are many technologies for leakage-prevention and plugging materials and drilling fluids, the existing materials and technologies have certain limitations, especially the difficulty of dealing with malignant lost circulation and longitudinal crack lost circulation during downhole construction operations, which is the most important factor that causes lost circulation in the various causes of downhole accidents. In the future the research and development or upgrade of leakage prevention and plugging drilling fluid materials and technologies should be strengthened to effectively reduce the occurrence of lost circulation [56].
2. Borehole instability under complex formation conditions

Especially in the process of drilling large sections of shale and complex structure wells, borehole instability easily occurs. How to design a reasonable drilling fluid density and how to improve the inhibitory performance of drilling fluids are the keys to effectively solving the problem of borehole instability under the geological conditions of complex deep wells, which needs to be carefully considered. The latest research results show that the multifield coupling method is required to solve the problem of borehole instability, especially the combination of rock mechanics and chemistry. Such a combination can achieve the purpose of well cementing. Although China University of Petroleum (Beijing) has established a new technology of wellbore-enhanced drilling fluid through the combination of internal and external factors and bionics methods, the drilling fluid's performance becomes worse when the temperature is higher than 200°C. Therefore it is necessary to develop well-cementing technology for the conditions of high temperature and complex deep wells.

7.2.1.3 The need to improve the mechanical penetration rate

Increasing the penetration rate is one of the effective ways to reduce drilling costs. Although water-based drilling fluid accelerators have been developed and certain effects have been observed in field applications, there are still the shortcomings, such as single product type and easy foaming. Actually, no mature speed-increasing water-based drilling fluid technology has ever been developed, resulting in the inability to significantly increase the penetration rate during the drilling of some complex wells, large-size well sections, and tight oil and gas. Therefore solving the problem of penetration rate is also a critical issue.

7.2.1.4 The need to enhance the high-temperature high-density water-based drilling fluid technology

Most of the remaining oil and gas in the world are buried in places that require deep and ultradeep wells, where temperature and pressure are high, and the drilling fluid is therefore required to have good resistance to high density and high temperature. For example, the temperature span in Songliao Basin and Tarim Basin is large, and the downhole temperature can reach above 250°C. The formation pressure is also very high, and large sections or even thick salt formations and salt-gypsum formations are often encountered. So the drilling fluids that are used in these areas must not only resist high temperature and density, but also have strong anticontamination ability. The current water-based drilling fluid technology has not made breakthrough progress in these aspects, which is one of the challenges facing drilling fluid technology.

7.2.1.5 The need to improve the performance of high-performance water-based drilling fluid technology

Although the advantages of water-based drilling fluids are prominent, there is still a certain gap in inhibitory performance and lubricity compared with oil-based drilling fluids. In the future the performance of water-based drilling fluids must be improved in these two aspects to equal that of oil-based drilling fluids. Although the China University of Petroleum (Beijing) has adopted bionic methods to develop bionic wellbore strengtheners and bonded lubricants so as to greatly improve the performance of water-based drilling fluids, there are still shortcomings in terms of general applicability and high temperature resistance.

7.2.1.6 Almost no achievements in the drilling fluid technology for ultradeep geothermal resources

The development and utilization of geothermal resources is one of the future development directions in the energy industry. High-temperature-resistant drilling fluid technology is one of the indispensable technologies for utilizing these resources. However, the current water-based, oil-based, and foam drilling fluid technologies can only meet the requirements of drilling relatively low-temperature geothermal wells. For example, the geothermal wells that are currently being developed overseas by the CNPC Great Wall Drilling Company have a maximum temperature of only about 250°C. Ordinary foam drilling fluids and even high-performance water-based drilling fluids can meet the needs of drilling in such case. With the development and utilization of hot rock, the drilling of stratum above 10,000 m, and the realization of the goal of transparent earth, the current drilling fluid technology is not sufficient to meet the requirements, and it is necessary to develop drilling fluid technology that can resist extremely high temperatures.

7.2.2 The development trends of bionic drilling fluid technologies

Effectively achieving safe, efficient, low-cost, and environmentally friendly exploration and development is the goal of the scientific and technological personnel engaged in the petroleum industry. Therefore the drilling fluid, which acts as the blood of drilling engineering, must be of high efficacy. With the increasing difficulty and complexity of drilling, higher requirements have been put forward for drilling fluid's borehole stability, lubricity, cuttings-carrying ability, protective effect on formations, and leakage prevention and plugging capabilities. The traditional theories and methods can hardly meet the needs of drilling in the new situation. So where can we get inspiration to help develop innovative and even original drilling fluid materials as the basis for new high-efficacy drilling fluid technologies to solve these technical problems?

The Chinese ancients said that learning comes from nature. The development of new bionic drilling fluid materials is an important source for solving the technical problems of drilling fluids to provide a new direction for realizing the sustainable development of drilling fluid technology. Based on learning from the natural world, introducing bionics into the field of drilling fluids and developing corresponding new materials for bionic drilling fluids constitute an effective way to achieve the above objectives.

Bionics is not only a means of innovation and entrepreneurship, but also a way to solve economic, environmental, and technical problems in a high-efficiency, low-consumption, green, and sustainable manner. The fusion of bionics and material science inspires people to develop new materials from the perspective of biomaterials. The use of biomaterials evolved over hundreds of millions of years as a model. Bionic simulations are carried out at different levels to provide new methods and new processes for developing and manufacturing new materials with the best performance, best environmental adaptation, and minimum energy consumption. The team led by Professor Guancheng Jiang of China University of Petroleum (Beijing) has invented a series of bionic drilling fluid materials and created the theories and technologies of bionic drilling fluids through learning from nature and has achieved good results. This is the first time that bionics has been introduced into the field of drilling fluids, so there are no examples and documents that can be used for reference in the world. Whether there are other biological principles and methods that could be more helpful in solving the international technical problems closely related to drilling fluids, such as well collapse and high frictional resistance, remains to be further explored. We should never stop learning from nature, and we must continue to explore the mysteries of nature, invent new bionic drilling fluid materials that can better solve technical problems, and establish corresponding new bionic drilling fluid technologies. Based on the forecast of the development trends of the international energy industry, we will briefly discuss the future development trends and key research directions of bionic drilling fluid technologies.

7.2.2.1 Improving the environmentally friendly and ecological properties of drilling fluid agents

Given the current lack of high-performance and low-cost environmentally friendly drilling fluid agents, especially the lack of diversified environmentally friendly agents, it is difficult to effectively improve the environmentally friendly fluid drilling technology, which will ultimately affect the implementation of drilling projects. Therefore the first priority should be developing environmentally friendly drilling fluid agents [57].

In the research and development of agents, imitating the ecological cycle chain mechanism or modifying natural and low-cost natural materials to invent bionic and environmentally friendly drilling fluid agents is an important research direction. In the near future it will be a promising way to modify natural polymer materials into environmentally friendly drilling fluid agents. Much attention should be paid to the development and utilization of natural agricultural by-products and processing by-products as well as optimizing and improving the performance of environmentally friendly drilling fluid agents, and improving the performance of natural polymer materials in terms of temperature resistance stability and inhibitory performance. At the same time, the selected inorganic salt must not only have no impact on the environment, but also have the characteristics of safety and nontoxicity [57]. In the future, attention should be paid to the development of agents with strong biodegradability, especially the application of nanotechnology, to better modify natural polymer materials. Therefore attention should be paid to the development of inorganic environmentally friendly drilling fluid materials.

7.2.2.2 Improving the resistance to high temperature and contamination of drilling fluid agents

The drilling of deep and ultradeep wells as well as the development and utilization of geothermal wells requires that the drilling fluid agents used should not fail in high-temperature and high-salinity environments. Furthermore, drilling fluid agents that are resistant to high cations can also reduce the activity of the drilling fluids, thereby improving the inhibitory performance of the drilling fluid system. Most of the current agents have the ability to resist temperature within 200°C. In terms of anticontamination ability there are some agents that resist salt saturation, but there are few agents that resist calcium ions or high-valent cations. It is difficult for calcium ions to reach 200 ppm, which makes it difficult for the performance of drilling fluids to meet the needs of drilling complex geological conditions. Therefore the research and development of agents with temperature resistance above 200°C and calcium ion saturation resistance or high-temperature-resistance and environmentally friendly agents that can replace sulfonated agents will have great development potential and prospects.

In the past 30 years, thermophilic organisms living in high-temperature environments have attracted the attention and interests of the scientific community. Especially since the discovery of thermophilic bacteria that can live at and above the boiling point of water, the research and application of thermophilic organisms have been further enhanced. For example, thermophilic bacteria can be used in genetic engineering and the fermentation industry and can also be used for bacterial leaching, oil and coal desulfurization, and anaerobic treatment of wastewater and waste [58].

Most organisms cannot survive in high-salt water. However, some organisms can survive in high-salt water or even water saturated with salt. For example, the Dead Sea is the saltiest water on the earth, with eight times salt content of ordinary seawater. However, there are some microorganisms such as bacteria and salt algae in the Dead Sea. Scientists have conducted studies on a type of bacteria in the Dead Sea, that is, halobacteria, and found that they can produce a protein to protect them from salt water. There is also a special kind of algae in the Dead Sea, that is,

Dunaliella salina, that can survive at a temperature of 53°C, reproduce at the low temperature of −27°C, and thrive in saturated salt water. Studies have found that *D. salina* has a salt-resistance protease in the body, which is a carbon dehydratase. Compared with the known carbon dehydratase (positive, negative, and neutral charges on the surface) in animals, the surface charge is all negative. The unique structure of the carbon dehydratase from *D. salina* has a special attraction for water molecules with a positive charge on one end and a negative charge on the other end. This structure can attract water molecules from the salty seawater, thus allowing it to survive in high-salt seawater [59].

Therefore by studying the mechanism of organisms living in high-temperature and high-salinity environments, we can imitate this mechanism to develop new drilling fluid materials with good resistance to high temperature and contamination.

7.2.2.3 Developing drilling fluid agents that can protect complex formations

At present, there are few types of drilling fluid agents that can protect formations, roughly including solid particles, such as ultrafine calcium carbonate, salt particles, oil-soluble resins, sulfonated asphalt; filming agents; and ideal filling agents. These protective materials have difficulty meeting the requirements of complex formations and special damage forms. Even the newly developed filming agent, biofilm, superhydrophobic agent, and superamphiphobic agent developed by China University of Petroleum (Beijing) have great potential to be further improved in terms of the efficiency and strength of the filming agent and biofilm, the degree of hydrophobicity and biphobicity of the superhydrophobic and superamphiphobic agents, and the temperature resistance, cost, and mechanism of these products. All the above factors seriously affect the effect of formation protection and require in-depth research over a long period of time to improve the protective effects or to develop new bionic formation protection materials.

Because the halophilic bacilli in the Dead Sea can produce a protein that protects them from salt water intrusion, these halophilic bacteria can survive in the Dead Sea. Therefore with the aid of the salt-resistance principle of halophilic bacilli, as well as the functions, structures, principles, and components of other organisms, new materials that similarly protect formations should be developed.

7.2.2.4 Developing intelligent nano drilling fluid agents

Intelligent technology is the development direction of the 21st century. The international community has aimed at the commanding heights of this technology and has decided to go deep into intelligent technology. However, the current international research on intelligent drilling fluid agents is still in its initial stage. In general, international research ideas intend to achieve intelligence through nanomaterials. Engaging in the research and development of intelligent nano drilling fluid agents is a new way to use bionics, but as yet, there have been no achievements in this field.

With the development of molecular biology and nanotechnology, bionics is developing in the direction of micro-nano structures and micro-nano system bionics and molecular bionics. On the geometric multidimensional scale of the research object, bionics can be divided into nanobionics, microbionics, macrobionics, and cosmobionics. Nanobionics is a new discipline created by the combination of nanotechnology and bionics. It involves imitating the functions and characteristics of organisms at the nanometer scale. Many surprising phenomena in nature are connected to the existence of nanounits in organisms. For example, natural spider silk has excellent characteristics such as light weight, high strength, and high toughness. Adding cellulose nanofiber materials to the copolymer mixture of ethylene oxide and epichlorohydrin can produce a product that can achieve the conversion between rigid and soft states, as in sea cucumbers, in a few seconds. This is a bionic nanocomposite material with self-adaptive, self-healing, and self-healing functions.

Therefore it is of great significance to use the theory of bionics; pay attention to learning from creatures in nature; focus on exploring the landscape, phenomena, principles, and laws of nature itself; carry out bionic research that mimics the habitat [60]; and develop nano drilling fluid agents that can solve the technical problems of drilling fluids, which will enable solving the problems faced in the drilling process and promote the development of drilling fluid technology.

7.2.2.5 Developing environmentally friendly and ecologically sound high-temperature-resistant oil-based drilling fluid agents

At present, the base oils that are used in oil-based drilling fluids are almost all toxic and do not easily degrade naturally, or the degradation rate is very slow. To deal with the problem of the serious environmental pollution of oil-based drilling fluids, it is necessary to develop environmentally friendly, naturally degradable oil-based drilling fluid agents. The first- and second-generation synthetic-based oils as well as the naturally degradable animal and vegetable oils that are currently being developed can be used in drilling fluids and meet the needs of drilling engineering, but they have some shortcomings, such as limited temperature resistance. For example, for naturally degraded animal and vegetable oil-based drilling fluids, the maximum temperature resistance is only 120°C, which limits the application of these fluids.

Nature is the source of life for human beings and other living creatures, and it provides humans with the abundant resources necessary for survival. Humanity has obtained the resources and wisdom for the development of civilization from nature, which has allowed the continuous evolution of human civilization and has brought about our modern progress and prosperity [60]. Therefore taking the natural environment as a habitat model and revealing the interaction mechanisms and change laws of different environmental factors will allow us to invent and create more bionic and environmentally friendly drilling fluid materials that meet the green and sustainable development needs of the natural environment, the human environment, and the drilling engineering environment [60].

Although the exploration, development, and utilization of petroleum and natural gas as energy sources have made great contributions to the progress of human society, the large amount of waste drilling fluid that is produced every year is increasingly destructive to the environment, and the consequences for the ecosystem are disastrous. Therefore human beings must consciously regulate and control the impact and solve the problem of damage to the ecosystem.

Nature is evolving all the time, and when people pay attention to the ingenious operation of nature, they can learn how to connect nutrients and energy, adopt a bionic strategy to provide energy with renewable resources, and turn trace resources into abundant resources. The human ecosystem has become more efficient [60]. Thus using the theoretical knowledge of bionics, the development of new bionic materials of high-temperature-resistant, environmentally friendly, and ecologically sound oil-based drilling fluid has solved some of the problems faced in drilling engineering and daily life.

7.2.2.6 Developing high-performance, environmentally friendly, and ecologically sound lubricants

With the expansion and deepening of oil and gas exploration and development, the increasing intensity of drilling and development, and the increasing difficulty of drilling, the complexity of drilling has become more prominent. Especially in the construction of directional wells and horizontal wells, as the inclination angle increases, the contact area between the drilling tool and the borehole increases, resulting in an increase in frictional resistance and torque, which greatly reduces the penetration rate and causes difficulty in the round trip [61]. In China, to solve the problem of lubrication and anti-pipe sticking, crude oil is usually mixed into water-based drilling fluids to reduce torque and frictional resistance. However, crude oil not only will interfere with fluorescent logging, but also can easily cause serious environmental pollution. However, there are few or almost no high-performance environmentally friendly lubricants at present. The lubricants that exist have serious adverse effects on the penetration rate, torque, and horizontal length of horizontal wells. Therefore the development of high-performance, environmentally friendly water-based drilling fluid lubricants has important practical significance [61].

During the drilling process, the movement of the drill bit and drilling tool in the stratum is similar to the shuttle of soil animals in the soil. To improve the penetration rate and wellbore quality, it is necessary to minimize the frictional resistance between the drilling tool, the borehole, and the drilling fluid; minimize the adhesion of the clay to the drill bit and drilling tool; minimize the energy loss resulting from friction; and maximize the mud pump power as much as possible to break the rock, especially in the process of drilling coalbed methane, shale oil and gas, tight oil and, and complex structure wells, which have higher requirement for the lubricity of drilling fluids.

For this reason it is necessary to add high-performance lubricants to the drilling fluid. The current lubricants are divided into solid particles and liquids. The working principle of solid particle lubricants is using the rolling of solid particles on the borehole. The working principle of liquid lubricants is almost all related to changing the characteristics and flow of the drilling fluid, and there is no bonding between the lubricant, the borehole, and the metal. However, the friction between the drilling tool, the drilling fluid, and the borehole is still large, the adhesion of the drill bit to the cuttings is strong, and the problem of balled bit is serious, which will have great negative effects on the rheological properties of the drilling fluids.

The shortcomings of liquid lubricants have been well solved by soil animals. Therefore it is important to introduce the principle of soil animals' reducing viscosity and resistance for developing high-efficiency lubricants that can bond with the borehole and the surface of the drilling tool. As a result, we can effectively control the inherent eddy currents in the flow interface, effectively reduce the friction in complex drilling constructions such as extended reach wells, increase the penetration rate, extend the life of drilling tools, reduce the operating costs, and improve the overall economic benefits of drilling.

7.2.2.7 Developing environmentally friendly and ecologically sound drilling fluid agents that resist ultralow temperature and acid gas

With the increasing efforts to explore the earth's oil and natural gas, people have found that there are abundant amounts of oil and natural gas resources in the extremely cold regions of the world. The drilling fluid agents that are required for the exploration and development of these resources must have the ability to withstand low temperature, ultralow temperature, and high temperature (during the drilling process, the temperature of the drilling fluid gradually rises after entering the formation), that is, the drilling fluids must have good performance under large temperature differences. At present, research on low-temperature and ultralow-temperature drilling fluid agents is rare.

The real threat to life from cold weather is not cold, but ice. Generally, cold-resistant organisms have a special cell membrane that can produce a protein antifreeze to ensure that they will not harden in extremely cold environments [62]. Discovering new animal species with a specific structure and new functions in such special environments will generate new and great appeal to bionics and bionics researchers.

Therefore the core technology of ultralow-temperature-resistant drilling fluid materials (mainly refers to the drilling fluid material in the aqueous state) is anti-icing. Inspired by the organisms that live in extremely cold areas, adding glycerol and other types of ethanol to the drilling fluid agent is one good way to prevent icing.

In recent years, with the increasingly in-depth study of organisms living in extreme environments, people have discovered many acid-resistant organisms in metal sulfur deposits, sulfur-containing hot springs and soil, acidic swamps, volcanic craters, and deep-sea sediments in the Pacific Ocean. Most of them live on inorganic nutrients and survive in extreme acidic environments without other life forms. For example, many acid-loving organisms live in the earth's hot acid hot springs, and they grow in an environment with a pH of 3.3–5.8 and a temperature of 55°C–75°C.

If we can determine the mechanism of their acid resistance, it will be of great help in the development of acid gas–resistant drilling fluid agents [63].

In addition, nature is a good teacher for mankind in terms of environmental protection and ecology. After hundreds of millions of years of evolution in nature, the energy use efficiency of each member in nature is much higher than that of human-made equipment with the same function and can create more value with less energy. Waste from one process always provides the nutrients, raw materials, or energy of another process, with no pollution or waste. Everything is a usable link in the natural ecological chain [64,65]. Therefore in the process of researching and developing drilling fluid agents, all resources are taken from nature. At the same time, after drilling is completed, the agents in the waste drilling fluid must be effectively integrated into the natural ecological chain and return to nature without destruction and harm, becoming an organic part of the natural ecosystem, thus forming a continuous cycle. This is the highest requirement for human beings to achieve environmental protection and ecology. In other words, we need to learn from the natural world. The whole process of manufacturing, use, and discharge of drilling fluid agents must reflect the protection of the environment and ecology and follow the principles of being green, ecological, sustainable, and best development in order to fully realize the optimal allocation and reasonable use of resources and energy.

Material production in nature has been sustainable for more than 3 billion years, while the process of human industrialization, which is not much more than 200 years old, has almost reached the point of unsustainable development. Only a small amount of the material obtained from nature in social production is used, and more than half of the material is thrown back to nature in the form of waste. This large-scale production based on high consumption, low output, and high pollution has exceeded nature's carrying capacity, and relying only on the treatment of pollutants to deal with the existing pollution is a very expensive procedure that has had little effect [63]. The harmless treatment of waste drilling fluids that has been widely carried out is one typical example (in fact, it cannot be harmless after treatment). In bionics the realization of the highest requirements for environmental protection and ecology put forward for drilling fluid agents is called sustainable bionics. It is an important means of solving existing resource and environmental problems and an important way to achieve the sustainable development of nature and humankind. Great importance should be attached to the continuity of bionic resources to ensure that all resources can be harmlessly returned to the natural environment after the entire life cycle of the product and that the returned resources can become ecological nutrients or be transformed into useful raw materials for other industrial production.

Sustainable bionics refers to making human design activities a positive factor for nature, economic and social health, and sustainable development while minimizing the negative impact of artificial products on the environment and ecology. Sustainable bionics requires the consideration of sustainability in every link from the generation of bionic concepts and the selection of bionic paths to the selection of bionic resources and the application and recycling of bionic products, promoting bionics to further incorporate nature. All resources are taken from nature and are returned to nature, achieving continuous and sustainable development in the cycle. That means that the use of raw materials in bionic drilling fluid agents is sustainable, the waste in the production process of drilling fluid agents is sustainable, and the drilling fluid agents that are used can be recycled and reused. Not only are the resources used in the whole process sustainable, but also the energy utilized should be clean and renewable. Relying on this kind of sustainable resources as well as clean and renewable energy can greatly reduce environmental and ecological pollution as well as alleviating energy shortages, exhaustion of resources, and other problems to truly achieve the design of sustainable resources and energy and returning to nature. Only by allowing natural resources and energy to operate efficiently, continuously, and cleanly can the healthy development of humankind be possible, allowing us to have a bright future.

To achieve the sustainable development of bionic drilling fluid agents, it is necessary to replace environmentally harmful and energy-consuming raw materials with nontoxic raw materials and clean energy to develop new drilling fluid agents. It is also important to properly plan recovery channels to ensure that the supply chain of bionic drilling fluid agents, the bionic drilling fluid agents themselves, and the recycling methods are all environmentally friendly and sustainable with the goal of guaranteeing the healthy, continuous, and efficient operation of humans and natural ecosystems.

7.2.2.8 Developing adaptive intelligent bionic drilling fluid agents

In the 21st century the important development trend of the petroleum industry is to incorporate intelligence, and drilling fluid agents are an indispensable part of the technological development of the petroleum industry. During the drilling process, from the surface to the bottom of a well, the nature of the rock encountered, the nature of the formation fluid, the distribution of formation pores and fractures, porosity, permeability, and water and oil saturation are constantly changing, vary greatly, and are heterogeneous, involving great uncertainties. The drilling fluids encounter different downhole environments with large changes and high frequency in terms of temperature, pressure, salinity, acid gas, pH, pores and fractures, and karst caves. Under these complex environmental conditions the drilling fluid must perform its due function, and when it encounters complex downhole conditions or accidents, it must be able to withstand these technical problems and ensure that the drilling process is safe, efficient, low-cost, and environmentally friendly. At present, the most commonly adopted method in the world is to add drilling fluid materials with different functions to the drilling fluid system, and when it encounters downhole complexity, materials that can handle the downhole complexity are added. This increases the cost of drilling fluid, and the effect is generally not ideal. Because the downhole conditions are changeable, these complex situations cannot be predicted accurately during the drilling process, and they cannot be grasped in detail and accurately once they occur, leading to a process of blindly taking treatment measures and often failing to solve the problems well. Therefore the best method is to use a drilling fluid that has a strong degree of adaptability, self-adaptability, and intelligence, which can better cope with damage to the performance of the drilling fluid caused by the changeable downhole environment. Furthermore, the drilling fluid that is used must have good self-repair ability, maintain relatively constant drilling fluid performance, give full play to the function of drilling fluid, and eliminate complex downhole conditions or accidents in the initial stage, completely avoiding the occurrences of complex downhole accidents such as lost circulation, well collapse, high frictional resistance, formation damage, and the formation of cuttings beds. In other words, the development of intelligent drilling fluid agents is an inevitable trend, and it is basically at a blank page at the moment.

The intelligence of drilling fluids is closely related to the self-adaptation, self-learning, self-organization, self-reproduction, biological sensing, and biological intelligence of natural organisms. The high degree of intelligence, high degree of coordination, and self-adaptability displayed by natural ecology is exactly what the self-adaptive intelligent bionic drilling fluid agent should emulate. Therefore to realize the intelligent function of drilling fluid agents, bionics is indispensable. Any creature has the ability to respond to the changing environmental conditions around it; that is, it has adaptive characteristics that are compatible with its living conditions. The life system is a very complex hierarchical structure system, with an automatic adjustment system featured by self-adaptation, self-learning, self-organization, self-reproduction, biological sensing, biological intelligence, and other functional characteristics. At the group level, individual level, cell level, and molecular level, life systems show certain strict and flexible automatic adjustment and self-adaptive characteristics. They obtain the greatest functional requirements with the lowest material and energy consumption and provide a natural biological blueprint for the innovation of drilling fluid technology and the manufacturing of intelligent drilling fluid materials. Therefore many aspects of biology are worth learning and imitating. The fine structure and excellent functional characteristics of biology can be transplanted to the field of drilling fluid technology for developing green, energy-saving, flexible, reliable, efficient, low-cost, and intelligent drilling fluid agents whose functional characteristics are close to biological ones. New intelligent drilling fluid materials are being developed that will be able to adapt according to demand. This has far-reaching significance that will benefit humankind [66,67].

In summary, from the analysis and exploration of the existing limitations and development trends of drilling fluid technology, it can be seen that the drilling depth and complexity continue to increase around the world, environmental protection regulations become stricter, and drilling costs are gradually increasing. In this situation the technical requirements and cost requirements for drilling fluids are also constantly increasing, but the global drilling fluid technology still has many limitations. Therefore research institutes and researchers should break the previous thinking model, broaden horizons, fully absorb fresh nutrients from other disciplines, jump out of the circle of drilling fluids to engage in the research of new drilling fluid materials and new technology, and strive to develop new drilling fluid technology as soon as possible, including original and disruptive technological achievements. Taking technological innovation as a breakthrough, we can solve the current and future technical problems of drilling fluids. While promoting the realization of drilling goals, we can realize the leapfrog development of drilling fluid technology, the sustainable development of petroleum industry and environmental protection, and the harmonious development of economic, ecological. and social benefits.

Innovation is the first driving force leading the development of science and technology. It is also an important force in promoting the development of the entire human world. Bionics is an important way to promote scientific and technological innovation. The initiative to take the path of bionic innovation is highly valued, which is particularly important for promoting the original innovation of drilling fluid technology. Among the major breakthroughs and innovations in various fields, many aspects are innovative applications of bionics and are closely related to bionics. Bionics is not a new science, but it is a field that humankind is increasingly pursuing in the quest for natural and technological innovation. Whether for basic research or cutting-edge technology, we find bionic innovation in a leading role.

In the fierce global technological competition, only innovators can advance, only innovators are strong, and only innovators will prevail. Bionic innovation technology is pushing the development of human society forward with an irresistible force. Researchers in bionics attach great importance to following the path of bionic innovation, which will take the lead in the fierce competition in the world and will never be defeated.

It is clear that the role of bionics technology in improving the original innovation of drilling fluid technology, key technology innovation, and systematic integration innovation will become increasingly prominent and will play an important leading role in the innovations of drilling fluid technology.

References

[1] W. Zhonghua, Review on the development of domestic drilling fluids and treatment agents, China Foreign Energy 18 (10) (2013) 34–43.
[2] M. Mingfu, Polymer drilling fluid, Pet. Nat. Gas Chem. Ind. (2)(1993) 36.
[3] W. Yuai, Polymer drilling fluid resistant to high salinity brine and its application, Drill. Prod. Technol. (3)(1993) 23–29.
[4] S. Engang, A new polymer drilling fluid and its application, Oilfield Chem. (4)(1989) 280–284.
[5] L. Yuqing, Research and application of cationic polymer drilling fluids, Nat. Gas Ind. 12 (3) (1992) 46–52.
[6] Zwitterionic polymer mud research experimental group, Research and application of zwitterionic polymer mud (1): zwitterionic polymer treatment agent and its mechanism of action, Drill. Fluids Completion Fluids 11 (4) (1994) 13–22.
[7] Zwitterionic polymer mud research experimental group, Research and application of zwitterionic polymer mud (2): zwitterionic polymer treatment agent and its mechanism of action, Drill. Fluids Completion Fluids 11 (5) (1994) 24–32.

[8] S. Bingzhong, J. Junbin, Research and application of composite metal ion polysulfonated oil-mixed drilling fluid system, Pet. Drill. Technol. 28 (5) (2000) 35–36.
[9] L. Jian, X. Xingjin, L. Pingya, et al., Research and application of amphoteric composite ion polymer mud treatment agent and mud system, Nat. Gas Ind. 11 (5) (1991) 42–49.
[10] B. Warren, P. van der Horst, W. Stewart, Application of amphoteric cellulose ethers in drilling fluid, in: SPE International Symposium on Oilfield Chemistry, Society of Petroleum Engineers, Houston, TX, February 5–7, 2003.
[11] L. Pingde, N. Yabin, Z. Mei, Development of amphoteric polymer drilling fluid system PL, Nat. Gas Ind. 25 (3) (2005) 71–73.
[12] X. Liqiong, Z. Jianchao, Application of zwitterionic polymer drilling fluid in Huabei Oilfield, Drill. Fluids Completion Fluids (1993).
[13] Z. Hao, B. Zhang, G. Xu, Applications of zwitterionic polymer HRH drilling fluid in Linpan Oilfield (PDF/HTML), Pet. Drill. Technol. (2) (2014) 57–63.
[14] F. Xuan, L. Yingwu, Q. Yulong, et al., Application of zwitterionic polymer drilling fluid system in Jilin DB block, Drill. Fluids Completion Fluids 22 (3) (2005) 75–76.
[15] M. Lal, Shale stability: drilling fluid interaction and shale strength, in: SPE Asia Pacific Oil and Gas Conference and Exhibition, Society of Petroleum Engineers, Jakarta, Indonesia, April 20–22, 1999.
[16] X. Yuzhi, L. Gongrang, L. Baofeng, et al., Application of super high density drilling fluid under ultra-high temperature on well Shengke-1, in: International Oil and Gas Conference and Exhibition in China, Society of Petroleum Engineers, Beijing, China, 8–10 June, 2010.
[17] J. Fuhua, Y. Zhigang, W. Hongxuan, et al., Turkmenistan ultra-high pressure huge salt-gypsum layer and malignant leaky pay zone drilling technology, Drill. Prod. Technol. 36 (4) (2013) 16–19.
[18] L. Xi, Research on Anti-Collapse Drilling Fluid Technology for Complex Mudstone Formations in Yuguo Block of Tuha Oilfield, China University of Petroleum, East China, 2013.
[19] J.C. Rojas, W.T. Daugherty, R.D. Irby, et al., New constant-rheological properties synthetic-based fluid reduces downhole losses in deepwater environments, in: SPE Annual Technical Conference and Exhibition, Society of Petroleum Engineers, Anaheim, CA, November 11–14, 2007.
[20] J.P. Plank, J.V. Hamberger, Field experience with a novel calcium-tolerant fluid-loss agent for drilling muds, in: European Petroleum Conference, Society of Petroleum Engineers, London, United Kingdom, October16–19, 1988.
[21] R.G. Bland, R.R. Waughman, P.G. Tomkins, et al., Water-based alternatives to oil-based muds: do they actually exist?, in: IADC/SPE Drilling Conference, Society of Petroleum Engineers, Dallas, TX, February 26–28, 2002.
[22] W. Dye, K. Daugereau, N. Hansen, et al., New water-based mud balances high-performance drilling and environmental compliance, in: SPE/IADC Drilling Conference, Society of Petroleum Engineers, Amsterdam, Netherlands, February 23–25, 2005.
[23] K. Morton, B. Bomar, M. Schiller, et al., Selection and evaluation criteria for high performance drilling fluids, in: SPE Annual Technical Conference and Exhibition, Dallas, TX, Society of Petroleum Engineers, October 9–12, 2005.
[24] H. Haoqing, Safe and environmentally friendly new water-based drilling fluid ULTRADRILL, Drill. Fluid Completion Fluid 21 (6) (2004) 4–7.
[25] A. Patel, E. Stamatakis, S. Young, et al., Advances in inhibitive water based drilling fluids—can they replace oil-based muds?, SPE106476, 2007.
[26] K. Qingming, C. Feng, et al., Application of ULTRADRIL water-based drilling fluid in Zhanghai 502FH well, Drill. Fluid Completion Fluid 23 (6) (2006) 71–76.

[27] W. Jianhua, Y. Jienian, D. Caiwei, Research progress of high-performance water-based drilling fluids, Drill. Fluids Completion Fluids 24 (1) (2007) 71–75.
[28] L. Ying, Research on Imitation Oil-Based Drilling Fluid System and Its Application in Mid-Shallow Horizontal Wells in Western Sichuan, Southwest Petroleum University, 2014.
[29] Schlumberger Service, Water-base drilling fluid systems [EB/OL]. <https://www.slb.com/resource-library/case-study/mi/hydraglyde-nam-cs>.
[30] Clear Solutions Service, Pure-Bore® [EB/OL]. <https://www.drilling-products.com/wp-content/uploads/2018/04/PDS-Pure-Bore%C2%AE.pdf>.
[31] Clear Solutions Service, Case Studies [EB/OL]. <https://www.drilling-products.com/pure-bore-use-demonstrates-significant-increase-in-oil-recovery-on-european-onshore-well/>.
[32] Clear Solutions Service, Case Studies [EB/OL]. <https://www.drilling-products.com/pure-bore-use-demonstrates-significant-increase-in-drilling-productivity-on-a-3285m-10775ft-tdv-shale-gas-well/>.
[33] M.A. Ramirez, D.K. Clapper, G. Sanchez, et al., Aluminum-based HPWBM successfully replaces oil-based mud to drill exploratory wells in an environmentally sensitive area, in: SPE Latin American and Caribbean Petroleum Engineering Conference, Society of Petroleum Engineers, Rio de Janeiro, Brazil, June 20–23, 2005.
[34] M.H. Hodder, A. Popplestone, P. Gwynne, et al., High-performance, water-based drilling fluid helps achieve early oil with lower capital expenditure, in: Offshore Europe, Society of Petroleum Engineers, Aberdeen, United Kingdom, September 6–9, 2005.
[35] M. Riley, S. Young, E. Stamatakis, et al., Wellbore stability in unconventional shales—the design of a nano-particle fluid, in: SPE Oil and Gas India Conference and Exhibition Society of Petroleum Engineers, Mumbai, India, 28–30 March, 2012.
[36] Bakerhughes Media Center, LATIDRILL water-based fluid system improves shale drilling [EB/OL]. <http://www.bakerhughes.com/news-and-media/media-center/product-announce-ments/houston-texas-august-7-2012-latidrill>.
[37] P. Yadav, B.A. Kosandar, P.B. Jadhav, et al., Customized high-performance water-based mud for unconventional reservoir drilling, in: SPE Middle East Oil & Gas Show and Conference, Society of Petroleum Engineers, Manama, Bahrain, March 8–11, 2015.
[38] Halliburton Service, Drilling fluids solutions [EB/OL]. <https://www.halliburton.com/content/dam/ps/public/bar/contents/Data_Sheets/web/Sales_Data_Sheets/H09805.pdf>.
[39] Halliburton Service, Drilling fluids solutions [EB/OL]. <https://www.halliburton.com/en-US/ps/baroid/fluid-services/drilling-fluids-solutions/water-based/performa-dril.html>.
[40] Halliburton Service, Drilling fluids solutions [EB/OL]. <https://www.halliburton.com/content/dam/ps/public/bar/contents/Case_Histories/web/PERFORMADRIL-Angola.pdf>.
[41] Z. Hanyi, Q. Zhengsong, H. Wei'an, et al., Evaluation and application of properties of polyamine high-performance water-based drilling fluids, Sci. Technol. Eng. 13 (10) (2013) 2803–2807.
[42] X. Mingbiao, Z. Chunyang, X. Botao, et al., Development of a new type of high-performance polyamine polymer drilling fluid, Nat. Gas Ind. 28 (12) (2008) 51–53.
[43] Y. Longbo, Research and Application of Organic Amine Drilling Fluid, China University of Petroleum, 2010.

[44] L. Kaihe, Q. Weigang, H. Liguo, et al., Research and application of multifunctional anti-high temperature and anti-collapse agent, Oilfield Chem. 29 (1) (2012) 10–13.
[45] L. Kaihe, H. Liguo, S. Tao, et al., Research on the influence of organic amine inhibitors on drilling fluid performance, Drill. Prod. Technol. 35 (2) (2012) 75–76. 96, 12.
[46] L. Kaihe, Q. Weigang, Z. Xiutai, et al., Research and application of mosaic shielding drilling fluid, Drill. Fluid Completion Fluid 29 (1) (2012) 23–26. 91.
[47] L. Kaihe, C. Yanan, Z. Jia, et al., Research and application of anti-collapse lubricant for drilling fluid, Offshore Oil 31 (3) (2011) 78–80.
[48] P. Jianzhong, L. Kaihe, W. Shuyong, et al., Research on polymer elastic microsphere reservoir protective agent, Drill. Fluids Completion Fluids 26 (2) (2009) 28–29. 131.
[49] L. Kaihe, Q. Zhengsong, W. Weidong, Research on application performance of polyhydroxy polymer anti-sloughing agent CXC-1, Oilfield Chem. (4)(2006) 293–296.
[50] W. Shuyong, Research and application of aluminum amine high-performance water-based drilling fluid, Drill. Fluid Completion Fluid 25 (4) (2006) 23–25.
[51] Z. Lihui, W. Zhijun, Z. Minli, Research and application of high-density organic salt drilling fluid for salt-gypsum formations, Drill. Fluids Completion Fluids 21 (4) (2004) 37–39.
[52] S. Kaijiao, X. Tongtai, Cesium formate/potassium solid-free drilling fluid and completion fluid research, Pet. Drill. Technol. 39 (2) (2011) 73–76.
[53] Z. Jie, S. Jinsheng, Y. Zhi, et al., Research on high temperature resistant solid-free drilling fluid, Pet. Drill. Prod. Technol. 33 (4) (2011) 45–47.
[54] Halliburton Service, Drilling fluids solutions [EB/OL]. <https://www.halliburton.com/content/dam/ps/public/bar/contents/Case_Histories/web/H010320.pdf>.
[55] Newpark Service, Water-based fluid system [EB/OL]. <https://www.newpark.com/evolution/>.
[56] D. Guangliang, The Status Quo, Challenge, Demand and Development Trend of Drilling Fluid Technology, Chemical Engineering and Equipment, 2017.
[57] H. Nianyi, Discussion on the Status and Development Trend of Environmentally Friendly Drilling Fluid Technology, Chemical Engineering and Equipment, 2017.
[58] Y. Xuezi, Work With Enthusiasm, Dedication, Love and Lubricate the Fruit-Liu Xiaoqing, Deputy Dean of Teaching at the School of Life Sciences, Capital Normal University, China Inventions and Patents, 2012.
[59] C. Xiaohong, Uncover the mystery of the existence of creatures in the Dead Sea, Ocean World (5)(1998) 24.
[60] L. Yunhong, R. Luquan, A preliminary study on natural habitat and its bionics, J. Jilin Univ. (Eng. Technol. Ed.) 46 (5) (2016) 1746–1756.
[61] X. Yang, Q. Xiaolin, L. Yongxue, et al., Research progress and development trend of water-based drilling fluid lubricants, Oilfield Chem. 34 (4) (2017) 721–726.
[62] C. Wenhai, Extremely addicted to life, Sci. Pop. (Middle Sch. Ed.) 11 (1) (2008) 2–4.
[63] M. Baoshan, Bionic design thinking and sustainable development, in: 2004 International Conference on Industrial Design Proceedings, 2004.
[64] W. Pei, G. Yanjie, J. Haman, translated Inspiration of Innovation: Inspiration and Creativity Inspired by Nature, CITIC Press, Beijing, 2015.
[65] J. Haman, The Shark's Paintbrush: Biomimicry and How Nature Is Inapiring Innovation, Nicholas Brealey, United States, 2013.
[66] S. Tibbits, 4D printing: making the world programmable, Glob. Sci. (2014).
[67] D. Jiahao, W. Xuan, 4D printing is a revolutionary technological breakthrough that will influence the development of the world's industry in the future, Sci. Technol. Rev. 31 (2013) 11.

Index

Note: Page numbers followed by "*f*" and "*t*" refer to figures and tables, respectively.

FSC
www.fsc.org
MIX
Papier | Fördert gute Waldnutzung
FSC® C083411

Kolibri 360 GmbH
Ferdinand-Jühlke-Straße 7
99095 Erfurt, Deutschland
produktsicherheit@kolibri360.de